IEEE ENGINEER

MW01602872

Volume VIII

WORKING IN A GLOBAL ENVIRONMENT:
UNDERSTANDING, COMMUNICATING, AND MANAGING TRANSNATIONALLY

by

Michael B. Goodman, Ph.D.
IEEE, Professional Communication Society
Senior Member, IEEE

The Institute of Electrical and Electronics Engineers, Inc.
New York, New York

IEEE Engineers Guide to Business Series

Editor:	Barbara Coburn
Typography:	m&m Creative Services
Cover Design:	David Beverage

Library of Congress Cataloging-in-Publication Data

Goodman, Michael B.
    Working in a global environment : understanding communicating, and
managing transnationally / by Michael B. Goodman.
        p.   cm. -- (IEEE engineer's guide to business ; v. 8)
    Innludes bibliographical references (p.).
    ISBN 0-7803-2263-0
    1. International business enterprises--Management.
2. Intercultural communication.   I. Title. II. Series.
HD62.4.G67 1995
658'.049--dc20
                                                    95-41904
                                                        CIP

Dedication:

To the memory of John Kailenta (1903-1995)
Businessman, Lions Club Chater Member,
Esperantist, World Traveler

Contents

PART III   UNDERSTANDING AND WORKING IN SPECIFIC REGIONS AND CULTURES

List of Illustrations and Figures

PREFACE

Growing up in Dallas I thought the world was a really big place. I was wrong.

Telecommunications, computer networks, and commercial jets bring the world to me every hour of every day. As a university professor and a consultant to business, my work has taken me to parts of the world I never dreamed of seeing. I count myself fortunate. Since you are reading this book and likely planning an overseas assignment, count yourself among the fortunate as well.

I have had a lot of help in writing this book. I am particularly grateful to my friends and colleagues at IEEE Professional Communicaiton Society: PCS President Deb Kizer of AT&T International and PCS past presidents Richard Robinson and Rudy Joenk, Herb Michaelson formerly of IBM, Scott Sanders at the University of New Mexico, Mike Markel at Boise State, Janet Robinson and David Milley of Lockheed/Martin, Sandy Friedman at Drexel University, Ron Blicq and Pam Kostur in Canada, Nancy Corbin of Westinghouse, and Ed Podell.

My good friends Dr. Henrich Lantsberg and Dr. Yuri Gornestaev of the Popov Society gave me a great deal of insight into the meaning of international communications. Thanks to the people I met in Moscow through them: Valery Zolotukhin and Aleksey Pavlov of the Russian Scientific-Technical Information Center; Drs. Sergei Marchenco and Yuri Savostitsky at the Institute for Automated Systems; Dr. Rostislav Vcherashny and Sergei Popov at the Russian Research and Information Institute for Industry (Informelectro); Dr. Valery Serov at the Humanitarium University; Pyotr Zoudkov of The Popov Society; Prof.

Anatoly Modin, Director of the Russian Academy of National Economy; Prof. Yuri Ukhin, State Committee for Science and Technology; Dr. Oleg Kedrovsky, President of the Information Workers' Association, and Svetlana Borisova of the State Committee on Science and Higher Education of Russia.

I am grateful to my colleagues at FDU's graduate school of business: Dean Paul Lerman, Walter Slegesky, and Jeanette Shurdrum. They gratiously shared their expertise and their experience with overseas programs.

Thanks to Joe Stampe and the alumni of FDU's Wroxton College who shared their overseas experiences with me, particularly Keld Alstrup of Volvo Cars of North America.

Thanks to Tony Petrocelli and Donald Kilpatrick of Harris Chemicals, Roger Leaf, John Cox, and John Keogel for discussions of global working experiences.

Thanks to the hundreds of engineers I have had the pleasure to have worked with at Grumman Corporation (now Northrop Grumman), Allied Signal, American Airlines, United Technologies - Norden (now Westinghouse), Bellcore, Exxon, Lawrence Livermore National Laboratory, Los Alamos National Laboratory, National Securities Clearing Corporation. Also, thanks to Emilio Rodriguez of Warner-Lambert.

Thanks to the faculty of Wroxton College, particularly Richard van Rees and Gerry O'Neill; and the numerous lecturers, speakers, Members of Parliament, journalists, and consultants who have come to our seminar at Wroxton College, including: Geoffrey Smith, Richard Clutterbuck of the University of Exeter, Philip Norton of the University of Hull, Bernard Corry of the University of London, Zerbanoo Gifford, Austin Mitchell MP, James Cran MP, Charles Kennedy MP, Michael Smith of Loughborough University, Anthony Meyer, Jerry Johns of the BBC, Michael Williams and Elizabeth Smith of the BBC World Service, Margaret Egan of McCann-Erickson Advertising Ltd., Sir Anthony Beaumont-Dark, Barbara Evans of AT&T (UK), Mike Judge and Phil Weare of Peugeot-Talbot, Nigel Holden of the University of Manchester, Graham Mole of Willis Corroon Group, Brendan Barber of the Trades Union Congress, Robbie Gilbert of the Congress of British Industry, Johnny Wright, Tony Keating and Tony Walford of UPS - Europe, Peter Milburn of Boots Healthcare International, Neil Spoonley, Mathew Cadbury of CadburySchweppes, Gerhard Kunz of the Federal Republic of Germany, Michael Snaith of Janssen Pharmaceuticals.

I would like to thank my colleagues at Fairleigh Dickinson University for their support: Mary Cross, Walter Cummins, Martin Green, Harry Keyishian, Walter Savage (Emeritus), Don Jugenheimer of the Communications Department; Robert Chell and Diane Wentworth of the Psychology Department; Peter Falley, Dean of the College of Arts and Sciences; Geoffrey Weinman, Vice President for Academic Affairs; Al Schielke, Asst. VP for Academic Affairs and Director of Overseas Programs: Francis Mertz, University President.

Also, thanks to the members of our board of corporate advisors for our graduate program in Corporate and Organizational Communication: Linn Weiss of Schering-Plough; David Powell and Dennis Signorovitch of Allied Signal; Gus Merkel of AT&T Company; Charlie Black of GPU Service Corp.; Hank Sandbach of Nabisco Brands; Dick Keelty of Warner-Lambert Company.

Thanks to our Schering-Plough Distinguished Professors: Tom Garbett, Corporate Consultant and Former Executive at Doyle Dane Bernbach; John Ryans of Kent State University; Sandy Sulcer and Cleve Langton of DDB Needham Worldwide.

I am deeply grateful for the research, critical evaluations, and comment of Pat Siccone, Victoria Rodriguez, and Natalie Vuksan - my fine graduate research assistants at Fairleigh Dickinson University; and former assistants Laura Hagen, Jill Reed, and Priyanka Kapoor; as well as our department secretary Chris Napolitano.

A special thanks to the graduate students I have had the pleasure to have worked and traveled with during our graduate seminars in international corporate communication and culture at Wroxton College. In particular, Thom O'Connor of Price Waterhouse and Kristi Roos of Maersk.

Special thanks goes to my colleague Nicholas Baldwin, Director of FDU's Wroxton College, England for constant conversations over the last several years about living and working overseas.

Thanks, also, to Barbara Coburn at IEEE for having asked me if I would like to write for engineers, and especially for her patience as I wrote this book.

And finally thanks to my wife Karen Goodman, my best critic, editor, and friend; and to my sons David Goodman and Craig Cook who take for granted that they are citizens of the world.

Michael B. Goodman
New York City, 1995

INTRODUCTION

"Act local, think global" has become the business mantra of the end of the century. The simplicity of the phrase can lure the unsuspecting into a simple-minded interpretation.

Much has been said, written, and videotaped on the need to compete in global markets. And much of what has been said and written about globalization of business emphasizes the notion that even though we may want a quick and easy method for entering markets outside our own country, the reality is that doing business in another country can be complex and difficult.

The complexity is in large measure cultural. We understand that working in a culture or nation different from our own forces us to master these determining forces to communicate and manage effectively. In addition to a familiarity with the history, the politics, the alliances and treaties, the art and literature of a country, an effective approach to learning about the transnational environment would include:

- Language
- Technology and the environment
- Social organization
- Contexts and face-saving
- Concepts of authority
- Body language and non-verbal communication
- Concepts of time

Language. Doing business successfully in an international, global, or transnational environment demands your attention to cultural, social, political, and religious practices, in addition to technical, business, legal, and financial.

Communication is key to each. Real communication—not just

cookbook dos and don'ts such as not showing the soles of your shoes in Saudi Arabia, or not shaking hands with a Japanese after putting something in your back pocket, or always finishing the bottle when a Russian begins to toast you, or not discussing business with a Mexican on the first business meeting. Such advice may be very interesting to read and think about, but it rarely recognizes that after the dos and don'ts run out, what do you do next. Such information is like having the pieces in a much larger puzzle, without a clear notion of the whole picture. That's where the "act local" part comes in. If you want to act local, you must be local.

In other words, understand the country you are doing business in. The first step is to make every effort to learn the language. Almost all nations - the French are a glaring exception - notice your effort to learn their language. This is more than just symbolic. Language encodes culture, and making an attempt to understand the words leads to trying to understand the way that people think.

The following examples of simple language differences are by now classics, almost cliches for international communication:
- General Motors' efforts to sell its Chevy *Nova* in Mexico. Nova sounds like the Spanish *No va*, or no go!
- And Ford's *Pinto* is Portuguese slang for a small male appendage.
- The popular *Bich* are Bic pens in the English speaking world for obvious reasons.

Technology and the Environment. The way people view technology and their environment is often culturally defined and can have an impact on international business communication. The way people view man-made work environments differs in perception of lighting, roominess, air temperature and humidity, access to electricity, telephones, and computers. People perceive their relationship to the physical environment differently. For some, nature is to be controlled, for others it is neutral or negative, and for others it is something for man to be in harmony with. Even climate, topography, and population density have an impact on the way people perceive of themselves, and that has a impact on the way they communicate, their concepts of mobility, and the way they carry on business.

Social Organization. Social organization, or the influence of shared actions and institutions on the behavior of the individual, has a strong impact on business communication worldwide. Institutions and structures tend to reinforce social values: the consensus of a group of people that a certain behavior has value.

For international business communication we might consider the following social structures which influence the workplace:
- kinship and family relationships;
- educational systems and ties to business;
- class and economic distinctions;
- religious, political, and legal systems;
- professional organizations and unions;
- gender stereotypes and roles;
- emphasis on the group or the individual;
- concepts of distance and attachment to the land;
- recreational activity.

Each one of these areas should be the focus of background research before going overseas. Some familiarity with the major works of art and literature will give you some insight into the social organization of the country you plan to visit on business.

Contexts and Face-Saving. Contexts and face-saving refer to the way one communicates and the situation in which the communication occurs. We refer to cultures that are high-context, like the Japanese, and low-context like the German. For example, a Japanese painting of a landscape will use only a brush stroke or two to represent a range of mountains, the details being left to the viewer to fill in. A high-context culture expresses figures with almost photographic detail. In a low-context culture like the British, details about class and education and even the place of birth are apparent in the clothes someone wears and the accent in their conversation.

Concepts of Authority. The concept of authority, influence, and power, as well as how power is exercised in the workplace, differs from culture to culture. For instance, in Western cultures such as the United States and Europe, power is the ability to make and act on decisions. Power for such cultures is an abstract ideal discussed and debated by philosophers and theorists from John Stuart Mill to Karl Marks. For Asian cultures, power and authority are almost the opposite of the Western concepts. Power results from social order. Asians accept decision-making by consensus, and decide to be part of the group rather than the leader. Understanding the concept of power helps shape a business communication strategy. The direct approach to communication, so effective in the United States, may prove crude and offensive in France or Japan.

Body Language and Nonverbal Communication. Body language and nonverbal communication are just as important in international and cross-cultural communications as they are in communications within a homogeneous culture. Watch movies and TV from a country you wish to visit before you go, as well as when you arrive. This gives you some cues to appropriate nonverbal behavior. Pay attention to kinesics (body movements), physical appearance and dress, eye contact, touching, proxemics (the space between people), and paralanguage (sounds and gestures used to communicate in place of words). Also colors, numbers and alphabets, symbols such as a national flag, and smell are important elements in international communication.

Concepts of Time. Concepts of time differ from culture to culture. In the twentieth century physicists such as Albert Einstein and more recently Steven Hawking have demonstrated that time in the physical sense is relative. For purposes of communication across cultures, it helps to consider time as a social variable. In the Caribbean, for example, the American tourist is frustrated to distraction when asking for a cab and getting the response, "Come soon." Time is defined culturally and by shared social experience.

COMMUNICATION IN THE NEW EUROPE

From the chaos and political instability that faced Europe at the end of the Second World War came a movement to unite the countries that led the world into global conflict twice in less than four decades. The concept was to link the countries economically in the hope that development of such ties would reduce the risk of to war. In addition, the goals of a European Community would contain nationalism which was the main cause of war in Europe, subdue a dominent Germany, create a barrier against Soviet Communism, strengthen prosperity at home and in world markets, and gain a strong European voice in international affairs. Since 1950, numerous treaties, agreements, and acts have evolved into the European Union, which as of 1995 was made up of Germany, France, Italy, Great Britain, The Netherlands, Denmark, Ireland, Belgium, Luxembourg, Spain, Portugal, Greece, Austria, Sweden, and Finland.

A main trading partner of the European Community is the European Free Trade Association (EFTA) which was formed in 1960 and includes Austria, Finland, Iceland, Liechtenstein, Norway, Sweden and Switzerland. Russia and the former Communist bloc countries of central and Eastern Europe — Hungary, Poland, Romania, Slovakia, Czech Republic, Bulgaria, and the countries of the Commonwealth of Independent States are developing new relationships with one another and with the European Union.

These enormous shifts in political and economic philosophy present a business communications challenge and opportunity. As barriers to trade are removed, the natural barriers of distance, culture, and language that have kept people apart for centuries begin again to play an important role in business transactions.

In the European Community, particularly since the fall of the Soviet Union, is a group of "Europeans." These are business professionals from all over Europe who tend to share a culture and belief system that has more in common with their international business counterparts in America or Asia. What they share with one another is often more than what they share with their own countrymen — taste in art, literature, music, recreational activities, cars, homes, attitudes towards work and money. What has emerged is a "global professional." For example, an advertising executive in France or England can function within the professional context almost anywhere in the world because of the commonality of activity. What has happened to almost all of the business professions is something that engineers have known and practiced for years — technical expertise translates well across many borders.

The business professional has emerged as a European class, often very well versed in the language and culture of the political nations he or she is working in and with. Other nations can hope to achieve this ideal of the international attitude and ability that Europeans have developed over centuries of trade.

COMMUNICATION AND THE PACIFIC RIM

For Americans doing business with nations of the Pacific Rim we can add to the difficulties of language and culture the added differences in context and face-saving. Context in communication usually refers to how much influence the situation exerts on meaning for the participants. Context can come from the impact of silence, from unspoken words, from inflection and tone of voice, from gestures, from timing of events, from form rather than substance.

Low-context cultures, such as in Germany and The United States, place a high emphasis on explicit communication, the law, and contracts. They rely on verbal communication, tolerate relatively little ambiguity, and place reduced emphasis on personal relationships and face-saving.

High-context cultures, such as in Japan and Latin America, place high emphasis on personal relationships, present information indirectly and often ambiguously or through nuance, and act at all times to preserve one's prestige or outward dignity — to save face. The word, laws, and contracts are seen as less important than the bindings of personal relationships.

In the high-context cultures of the Pacific Rim, business communicators from low-context cultures such as the United States will be confronted with controlled use of silence, or communication through intuition. The Japanese have elevated such meaningful silences to an art form and call it *haragei. Hara,* or literally belly, is the English equivalent of heart or center of one's being, the center of feelings, courage, and understanding, as well as the wisdom gained through one's experience. *Haragei* is the opposite of the argument or verbal confrontation so common to the business communication of Westerners.

Another concept in the high-context cultures of the Pacific Rim is the Korean *kibun* or moods or feelings. Koreans are very sensitive to maintaining harmony and go to what Westerners consider great lengths to maintain their own *kibun* as well as everyone else's. The concept plays a role in the aversion of most Asian Pacific Rim nations to bring bad or unpleasant news. It is also related to an unwillingness to say "no" directly as a way to save face.

In high-context cultures the differences between the surface truth and reality may be much more important than in low-context cultures which often make no such distinction. The Japanese use the terms *tatemae* and *honne. Tatemae* is the facade of a structure like a building, and *honne* is one's true voice, what one really thinks and feels. Every culture has such concepts to some degree. Even a quick scan of most European novels of the last century, or the works of American novelist Henry James, reveals the richness that exists in the difference between the public expression and the private thoughts of individuals.

The status of the Pacific Rim as an economic and political force requires corporations of any size to develop a business and communication strategy that meets the business challenge effectively. Making a strong and conscious effort to understand the concepts of contexting and face-saving is essential for any effective assignment in almost every nation of the Asian Pacific Rim.

COMMUNICATION WITH DEVELOPING COUNTRIES

Americans doing business in developing countries should make every effort to understand the cultures, customs, and language of the people they communicate with in those countries or regions. While working and talking to people in developing countries, make no excuses about being from another culture. Chances are they know a lot more about you from movies, books, and mass media than you know of them. Many of the business professionals were educated in the United States and are more likely of a social class higher than most of their countrymen.

Developing countries may appear to have disadvantages compared to some of the advanced economies of the world. But remember, they often have a rich artistic, religious, and cultural heritage that should be the focus of your building a business relationship with them. The economy of many developing countries may be built on many families or sole proprietor companies. They can therefore compete in a global economy because in their small companies the flexibility and lack of bureaucracy gives them a comparative advantage.

Since they might also know English, you may show your interest in them by at least reading their literature in translation, being aware of their cultural and artistic accomplishments, and making an effort to learn their language.

But also be proud of who you are. Nothing seems less genuine than a foreigner who seems to "go native" at the expense of his or her own culture.

To communicate in a global environment, the understanding of contexts, situations, languages, cultures, and motives should prove an appropriate and valuable approach to almost any new culture. In short, make every effort to understand your audience's needs and expectations.

COMMUNICATION TECHNOLOGIES OVERCOME BARRIERS OF TIME AND SPACE

Working in a global environment underscores the importance of some communication media. Technologies such as satellites and e-mail have more and more replaced the telex, the fax machine, and the telephone in international business.

It is common in some technology based companies to have groups all over the world work on projects around the clock. These professionals and technicians are connected to one another by computer networks. For example in New York, a group will work on a project. At the end of the work day, they will hand off the job to another group on the computer network in Los Angeles. In this way the work is passed around the world and around the clock, overcoming the communication barriers of time and space.

Technology advances in communication have created the global business environment that challenges us today.

WHAT TO EXPECT IN THIS BOOK: AN OVERVIEW

The book that follows goes into each of the topics in more depth. There are three main sections, and sixteen main areas are covered. (See Table of Contents)

The **Further Reading** section lists the books used as sources. These offer a solid foundation for making a broader or deeper study of the global workplace.

The seven **Appendices** offer:
- a model of a corporation's code of international business ethics
- a set of actions to help you "act local, think global"
- a table of metaphors for national cultures
- a matrix for corporate cultures

- a worksheet for gathering facts for a country analysis
- a checklist to help you plan as your prepare for your overseas assignment
- a list of the sources of help you can count on from the United States Mission when you are in a foreign county

PART I
UNDERSTANDING THE GLOBAL BUSINESS PROCESS, ITS ENVIRONMENTS, AND CULTURES

- Language and Communication Issues in Transnational Business
- Social Organization and Working Transnationally
- Understanding the Impact of Contexting and Face-Saving in a Global Environment
- Understanding the Impact of Technology and the Environment on Working Transnationally
- Understanding Concepts of Power, Influence and Authority in a Global Environment
- Body Language and Nonverbal Communication for Global Business
- Concepts of Time and Communication in an International Environment

CHAPTER 1
Language and Communication Issues in Transnational Business

As mentioned in the introduction, doing business successfully in a transnational environment demands your attention to cultural, social, political, and religious practices, in addition to technical, business, legal, and financial. Communication is key to each. If you want to communicate effectively, to "act local," you must be local. English has become the international language of air traffic, of radio transmissions at sea, and of science, technology, and engineering. The use and importance of English as the international language of business often creates a serious handicap for native speakers since it lulls them into the belief that they can compete on a global scale with only knowledge of their own language.

To be local means that you must make every effort to learn the language. Almost all nations, as mentioned, notice and appreciate your effort to learn their language. This is more than just symbolic. Any anthropologist will tell you that the language of any group of people encodes its culture. We understand that all disciplines of engineering have a specialized language. Often that language, for example of Information Technology or Artificial Intelligence, can be a common bond between you and the people you are working with in another country. English speaking people often take for granted that English has become the international language of science, engineering, aviation, and business practices and procedures. But working in a foreign country also means dealing with people who do not share that bond of a common business language. Learning their language, or making the effort to try, is a major step toward understanding how your hosts concieve the world and how they think.

Learning intercultural communication is a three-step process: awareness, knowledge, and skills. The first step, awareness, means one recognizes that everyone is born into and brought up with a particular way of viewing the world, and that others brought up in a different environment have, naturally, a different world view. By seeing the relative nature of interactions with people of other cultures, the motivation to learn more about them should follow. Knowledge of their symbols, their history, their beliefs, their values, their literature, their customs and ceremonies will result. Understanding their approach to the world is enough. No one expects you to embrace another's beliefs, but a firm intellectual understanding of how values differ from country to country is essential to working with others in a multicultural, global environment. Mastering this skill goes beyond awareness and knowledge by putting understanding into practice. And to attain skill at intercultural understanding is to practice recognizing the symbols, heroes, and rituals of others.

However, some people are not suited for such international, global environments. The following personality traits are often considered symptoms of an individual who is a poor risk to work in a foreign country:
- ego uncertainty, either overly inflated or too self-effacing
- inability to adapt to climatic conditions
- inability to tolerate change or uncertainty
- emotional instability
- lack of patience to work and operate under foreign conditions
- racial or gender intolerance
- extreme political views, either to the right or left (conservative or liberal)
- extreme religious views
- extreme class prejudice
- loners, unable to work effectively in teams or groups

Most companies offer a briefing seminar for the people they are sending to work in a foreign country. This may include the country's history, geography, customs, health and hygiene practices, dos and don'ts, as well as what to bring. Such briefings can take a few days to several weeks.

However, as advocated, the best way to prepare is to learn the local language. This takes time, even for the gifted. To feel confident that you can function in another country may take a few months of full-time effort. If you are learning the language in the country itself, it should take less time because you are immersed in the language and culture. Most companies do not allow enough time for their managers to learn the language, but they should. They should also make provisions for the members of the family to learn, if they will be going too. In the last few years many companies have gotten around the practice of language and culture training by posting their managers for only a few months in the foreign country, and leaving the employee's family in the home country. The argument is that such employees never "go native" nor do they ever lose sight of the home country perspective. Such an approach has merit, but it is not an argument for or against learning the language. As we will see in the following chapters, in most countries doing business transnationally is not a 9 to 5 affair. You must also live there too — shop, eat, travel, learn, have fun.

Let me illustrate the point about the importance of understanding and acting upon the differences in language, and, by extension, the importance of learning the language of the other country. The following is a brief comparison of the way Americans express themselves, as opposed to the British. Since both speak English, this illustrates how much is revealed in the different ways people express themselves.

Let's begin with greeting rituals. Americans walk right up to someone, hold out their hand and say, *Hi, I'm John,* or *Hello.* An English business professional, male or female, would wait to be properly introduced by a third party before saying anything. I was at a public relations reception for the BBC for almost half an hour before the public relations director was brought over to me and introduced and we began talking. He knew I was there, but he needed someone to introduce us.

Other examples: while driving in England you *Give Way,* in the United States you *Stop;* in England you *join the queue* and never *jump the queue;* in the United States you *line up* and lots of people *butt in line.*

OK, you say. You are convinced that language is an essential skill in the transaction of international business, but like almost all Americans, and like almost all American corporations, you have not mastered another language. That is why you choose a career in technology in the first place. Your strength is with numbers and the qualitative side of business, not with languages. What can you do to overcome language as a barrier to international business communication? David Victor (*International Business Communication*: 38-45) suggests five steps that limit the risk of misunderstanding and overcome the challenges which language can create in business communication. They are:
- adjust untranslated communication
- carefully select translators and interpreters
- personally review translated documents
- pay attention to names and key terms
- use back-translation

Untranslated Communication: Often in conducting business in a foreign country you do not speak their language, but they have learned English as a second language. Even if you are conducting business in your primary language without an interpreter because the other party speaks your language, you can make things easier:
- First, make an effort to use simple language free of idioms, slang, and colloquial terms. Such words and phrases, while clear and useful for native speakers, are a source of confusion for others. If you like an idea, say so rather than describing it as a *touchdown* or a *home run.*
- Second, keep the words you choose simple. Be careful to determine the level of expertise of the listener or reader, but don't assume that because they may have limited vocabulary in you language that you must also simplify the content. Speak slowly. When you slow

down it allows the listener the time to translate.
- Third, rephrase or repeat key words and phrases. This gives your listener several chances to grasp your exact meaning.
- Fourth, bring written versions along with you to support your conversation. Most speakers of a foreign language have far better reading than listening or speaking skills. This allows them to translate difficult or complex information later. This is particularly true of engineering and technical information which may be difficult enough for a native speaker to understand in the primary language.
- Fifth, become aware of and try to use cognates, words that have similar roots and may be identical in both languages, in your dealings. For example, in German you use a written *kontrakt* to formalize the deal, and then you might celebrate by drinking *bier.*
- And finally, stop often to summarize. This practice allows you to make clear what you wanted to say, as well as to express your understanding of what the other person meant.

Selection of Translators and Interpreters: In most situations you can carry on quite well without the use of a formal translator. However, if you find that the information is extemely important or sensitive, or both parties involved do not speak one another's language, then arrange for a translator or interpreter. Look for a person who is reputable, usually from an established translation service; familiar with the dialect and culture of the group you are dealing with; and has an understanding of and expertise in business terms and strategies. When you are using an interpreter, remember that you are having a conversation with your managerial counterpart, not the translator, so maintain eye contact with that person. Also speak slowly enough so the interpreter can communicate accurately.

Personal Review of Translated Documents: Even though you do not know the language of the translation, you can still help out in areas that you are familiar with and check the translation for errors in: spelling of personal names, company names, brands, trademarks; the physical appearance and neatness of the document; accent marks and other symbols; mathematical equations; statistical and tabular representations; graphs, charts, diagrams and other figures and exhibits; numerical conversions such as US measures to the metric system.

Attention to Names and Key Terms: Pay particular attention to a person's name and make every effort to pronounce and spell it correctly. Also, be sure to use the correct company name and spelling.

Back-Translation: This two-step process has one translator make the translation into the target language; another translator takes that document and translates it back into the original language. Then the two are compared for possible errors or misinterpretations.

CHAPTER 2
Social Organization and
Working Transnationally

Social Organization is overt external behavior that is easy to observe. For most foreigners the differences are apparent — dress, facial expression, greeting rituals. Though it is easy to see the differences, the social forces are difficult for members of the society to overcome, and the influence of shared actions and institutions on the behavior of the individual has a strong impact on business communication worldwide. Institutions and structures tend to reinforce social values, that is the consensus of a group of people that a certain behavior has value. These institutions, because they are different the world over, are by nature artificial. When working globally, it helps to keep in mind that the imposed and artificial nature of social stuctures do not in any way mean the people are less committed to them. In fact, they may be even more committed.

To work successfully in an international business environment one should consider the following social structures which influence the workplace:
- kinship and family relationships;
- educational systems and ties to business;
- class and economic distinctions;
- religious, political, and legal systems;
- professional organizations and unions;
- gender stereotypes and roles;
- emphasis on the group or the individual;
- concepts of distance and attachment to the land;
- recreational activity.

Each one of these areas should be the focus of background research before going overseas. (See Appendix 5) Some familiarity with the major works of art and literature will give you some insight into the social organization of the country you plan to visit.

Kinship and Family Relationships. Americans, particularly technical professionals, see work and professional activity based on merit, performance, and ability. Americans, along with several Western European countries (Fig. 2.1), have little understanding of the kind of relationships found in much of the world. These relationships place family ties and kinship above performance. Such differences raise issues about hiring, firing, nepotism, securing loans, trust, negotiations, and cultural differences.

In international business, be aware of the family ties among individuals, and also recognize such relations as the social bond of trust that extends into business dealings. As an expert or manager with a background in technology, be aware of the strength of family relationships. Such bonds are strong and

WEAK TIES BETWEEN KINSHIP AND BUSINESS	**STRONG** TIES BETWEEN KINSHIP AND BUSINESS
Australian British Canadian Danish German Icelandic Norwegian Swedish United States	African Arabic Chinese Greek Indian Indonesian Iranian Italian Korean Latin American Portuguese Spanish Turkish

FIGURE 2.1 Kinship Ties in Several Cultural Environments
Source: Victor, International Business Communication

are a common part of organizational stucture in Asian countries, Africa, Latin America, and much of Europe. People from countries with weak kinship ties consider this a display of nepotism which is often forbidden by law. A misunderstanding of strong family ties as favoritism or nepotism can present a major obstacle to working with the people in your host country.

Educational Systems and Ties to Business. Like wealth and power, education is closely allied to the haves and have nots in most countries and cultures around the world. Strong educational ties to business, an educational elite, exist in much of industrialized Europe, Asia, and Russia. To understand the educational structure of a given country, think about who has access to the education, the links the educational institution provides to business for the members of the institution and the type and styles of teaching.

Access to education in the industrialized countries takes four major forms. In Great Britain and most of Europe education is on a two-tier or two-track system. Some students are put on a track that leads to the universities, and others are placed on work or trade-related tracks. Tests early in the student's life — at about age 12 or 13, and then again at about 17 — determine this track.

In Japan and the United States access to education, particularly higher education is based on testing, with a very high degree of open access. Japan, more than the U.S., has strong links between attending the right university and social mobility. If a Japanese attends Toyko University, for example, that person is set for life. That may also have been true in the United States thirty or forty years ago if you attended MIT, Yale, or Harvard, but it is less likely to be perceived as the ticket to success it once was. In Russia the system has much more open access, but those who are successful become an educational elite. In all the industrialized countries, education is a form of power and influence. In democracies it is also a form of social mobility.

The nature and style of education has a great deal to do with the way the people of that culture frame the world. In Great Britain, the educational elite learn in a humanities-based environment, and shun the practical applications of thought and embrace learning for its own sake. If the emphasis is on technological issues, a country is much more likely to be open to business ideas because much of the business activity is related to advances in technology. The more you know about how the people of the country acquire their knowledge and then transmit that information to their colleagues, the better prepared you will be to work with them effectively.

Class and Economic Distinctions. Social class in many countries is closely liked to education. In Great Britain the upper classes strive for government posts in service to their country, or they try work in a non-business context. They associate business with the trades. Class in the United Kingdom, Middle Eastern countries, and in South America is also related to differences in wealth, as well as family ties. In egalitarian cultures such as the United States, wealth is valued when it is associated with achievement, rather than inheritance. The self-made man in the United States is almost universally revered. In the United Kingdom the opposite is true — the wealthy, self-made man is often seen as a social climber. Such class distinctions cast a stigma on such audaciousness, and produce no animosity toward the idle rich.

Religion also has a connection to power and class. In Catholic countries the church wields enormous power. In India the class structure is connected to the social structure of the caste system.

Religious, Political, and Legal Systems. Some countries have a state religion. The state religion in Great Britain is the Church of England. This has little to do with religious commitment and belief among the subjects of the Queen. For example, an Englishman is appalled at the emotional debate and violence over the abortion issue in the United States because it is not the business of the church to legislate behavior, or for the state to invade the privacy of its people. The place of religion as an instrument of the state makes it political. Even with a state religion, church attendance in England is among the lowest in the Western world.

Politics in much of Europe is intertwined with business because the governments there matured as the centers of power before businesses did. Power is vested in a parliamentary body. All decisions of any consequence are made in that political forum. Americans are often daunted by the power of the state in Europe to complicate even the most routine business transaction. There, as elsewhere, nothing gets done without the knowledge and consent of the central political body.

By contrast, the use of the legal system is not nearly as common as in the United States. Most Members of Parliament, in England for example, are not lawyers. They were trained in the humanistic mode, so their goal is to promote the best interest of the nation. European democracies also will be more inclined to solve problems with an approach that is more socially inclusive. Such nations often interpret democracy in a more socialistic way than is done in the United States.

Professional Organizations and Unions. When managing technical projects in countries that have a strong democratic, egalitarian tradition, organizations exercise a great deal of power over technology. Such countries might be perceived as technocracies or social structures based on technical performance. Technical expertise in even a highly class conscious country can be a pass into the ruling class. After all, techology is a path to power.

In some countries power was and is concentrated in the unions to balance the power elite of the ruling classes.

Gender Stereotypes and Roles. Recently, Deborah Tannen in her books *You Just Don't Understand* and *Talking 9-5* explored the role of gender in business in the United States. Some have criticized her analysis of "power talk" as a matter of socialization in the world of work on the part of some women.

No matter what side of the debate you are on, when you are overseas, the roles of men and women at work are all too clear to even the most myopic manager. For instance, on a trip to Russia, I was taken to numerous headquarter buildings of various agencies that were responsible for computing equipment and electronics, as well as research and development. In those offices, which have some similarities to high-tech engineering firms in the United States, it was very clear that the older directors and managers expected their female assistants to serve coffee and cookies. As an American I was a bit uncomfortable with this gender role activity since it has faded rapidly from the American workplace. However, this "men's club environment" is still prevalent in high-tech organizations.

Also, in places such as the Arab Middle East, women managers and technical experts, as well as teachers and administrators who are single, find themselves in difficult, if not unbearably uncomfortable circumstances. To be a single professional woman in that context is to be shunned and avoided, segregated and ignored.

In addition to these professional gender-role stereotypes are the gender biases that members of a culture acquire just by being raised in that region or country. Geert Hofstede, in his *Cultures and Organizations: Intercultural Cooperation and Its Importance for Survival,* explores the concepts of masculinity and femininity in different regions of the world. (See Figure 2.2) His interpretation of masculinity and femininity is linked with emotional closeness and empathy. He looked at behavior in over 50 countries, and did not differentiate between the behaviors of men or women. In other words, he was looking at these cultures for behavior that was not linked to gender. As a manager in foreign countries, knowing the behavior style is more important than perceiving that style to a gender-linked behavior. As the attitude toward women in international business is more and more driven by the attitudes and behaviors in the United States and Europe, it might be more useful to see the styles in other countries.

Nancy Piet-Pelon and Barbara Hornby in their *Women's*

SCORE/RANK	COUNTRY or REGION	MASCULINITY INDEX
1	Japan	95
2	Austria	79
3	Venezuela	73
4/5	Italy & Switzerland	70
6	Mexico	69
7/8	Ireland & Jamaica	68
9/10	Great Britain & Germany	66
11/12	Philippines & Columbia	64
13/14	South America & Equador	63
15	U.S.A	62
16	Australia	61
17	New Zealand	58
18/19	Greece & Hong Kong	57
20/21	Argentina & India	56
22	Belgium	54
23	Arab Countries	53
24	Canada & Malaysia	52
25/26	Pakistan	50
27	Brazil	49
28	Singapore	48
29	Israel	47
30/31	Indonesia & West Africa	46
32/33	Turkey & Taiwan	45
34	Panama	44
35/36	Iran & France	43
37/38	Spain & Peru	42
39	East Africa	41
40	Salvador	40
41	South Korea	39
42	Uruguay	38
43	Guatemala	37
44	Thailand	34
45	Portugal	31
46	Chile	28
47	Finland	26
48/49	Yugoslavia & Costa Rica	21
50	Denmark	16
51	Netherlands	14
52	Norway	8
53	Sweden	5

FIGURE 2.2 Hofstede's Masculinity Index for 50 Countries and 3 Regions

Source: Hofstede

Guide to Overseas Living offer this anecdote for the professional woman posted overseas:

> One of the most serious concerns of women who work overseas is professional acceptance. While in most societies women can be found in such areas as teaching, nursing or office work, the expatriate woman working in other fields will find professional women on an equal level in only a few countries. Very often she will find that she does not work with women at all. In such cases, her presence may so disorient her colleagues that they literally cannot "hear" her when she expresses her ideas, responding only to the strangeness of encountering a woman in a professional or executive role. The professional woman must develop a tough skin, be willing to move slowly and cautiously, and have both great self-control and self-confiennce. (Piet-Pelon and Hornby, 105)

Emphasis on the Group or the Individual. In addition to notions of masculinity and femininity, cultures have differing ideas of the individual and the collective society. Notions of *I*, *we*, and *they* are culturally defined and can help in understanding the culture.

Once again Hofstede has given us a fine working definition: "Individualism pertains to societies in which the ties between individuals are loose: everyone is expected to look after himself or herself and his or her immediate family. Collectivism as its opposite pertains to societies in which people from birth onwards are integrated into strong, cohesive groups, which throughout people's lifetime continue to protect them in exchange for unquestioning loyalty." (Hofstede, 51)

Hofstede offers an Individualism Index that should help place a given country or region along a continuum of individualism and collectivism. (See Figure 2.3)

The United States is of course a nation that embodies indi-

SCORE/RANK	COUNTRY or REGION	INDIVIDUALISM
1	USA	91
2	Austria	90
3	Great Britain	89
4/5	Canada	80
4/5	Netherlands	80
7	Italy	76
8	Belgium	75
9	Denmark	74
10/11	Sweden	71
15	Germany	67
20	Spain	51
22/23	Japan	46
22/23	Argentina	46
26/27	Arab countries	38
30	Greece	35
32	Mexico	30
37	Hong Kong	25
43	South Korea	18
44	Taiwan	17
50	Venezuela	12

FIGURE 2.3 Selected Individualism Index Values for 50
                   Countries and 3 regions

Source: Hofstede

vidualism. Independence, self-reliance, and freedom are at the core of defining for Americans what it means to be an American. In Japan, by contrast, the saying "The nail that stands up will be hammered down," should give anyone working in that country an indication that the individual is subordinate to the group or organization.

Concepts of Distance & Attachment to the Land. How people view the concepts of distance and their attachment to where they live is also related to the social and cultural make-up of the country. A link exists between a nation's mobility and how long it has been settled. For instance, the percentage of people relocating in France and Great Britain is approximately 10%; compared with almost 20% in "newly settled" New Zealand, Canada, and the United States.

The mobility of a culture can be thought of as high, static, or phasic. In high mobility cultures the language emphasizes movement and is focused on verbs — *moving up, on the go, fast track, rising star*. Static mobility cultures place heavy emphasis on the attachment to one's place of birth or home region. Here, the attitude toward strangers may be less than understanding because the local inhabitants themselves have little belief in the value of leaving home and would wonder about someone from another country who did.

Phasic mobility is found where people are willing to move for a limited period of time. In a global economy we find the proliferation of *Gastarbeiters* — Turks and Slavs in Germany, Algerians in France, Mexicans in the United States. These are foreign nationals who come to countries with a higher standard of living than their own. They work, sometimes for decades, with no effort to become citizens. The host countries either treat them as foreigners as in Germany and France; as equal to its own citizens as in Sweden; or as "minority natives" as in Saudi Arabia.

It is important to know these subtle differences in other countries because managing technical projects often brings you in contact with the full range of professionals, administrators, and skilled and nonskilled labor.

Recreational Activity.

CRICKET

as explained to a foreign visitor

You have two sides one out in the field and one in.

Each man that's in the side that's in goes out and when he's out he comes in and the next man goes in until he's out.

When they are all out the side that's out comes in and the side that's been in goes out and tries to get those coming in out.

Sometimes you get men still in and not out.

When both sides have been in and out including the not outs

That's the end of the game

HOWZAT!

Source: Marylebone Cricket Club Banner

In Great Britain, cricket and football are sacred. Tennis is a religious experience; gardening and DIY (Do It Yourself) are of higher importance than everything except dogs. Knowing the ways a culture enjoys itself will help you understand that culture.

In Japan, writing Haiku is a form of recreation that may be lost on Americans, particularly in the context of the Karaoke Bar and the explicitly adult content of Japanese comic books. Haiku is a lyric form of poetry containing seventeen syllables in three line of 5,7,5 syllables begun in the 16th century. Haiku must adhere to strict form and use of only natural images. Literary tradition, Buddhism, and Taoism add symbolic power to the natural images. Almost every businessman in Japan writes Haiku, and has been to the Karaoke bar and reads comics.

CHAPTER 3
Understanding the Impact of Contexting and Face-Saving in a Global Environment

Contexting and *face-saving* refer to the way one communicates and the situation in which the communication occurs. All communication, global or otherwise, occurs in a context. The more two people share knowledge and experience, the less important it is for them to express directly what they wish to say or write. The less they share, of course, the more they must express in words and gestures to be understood. This is the concept of contexting, which should be considered as a learned cultural behavior.

Contexting: Members of the same culture share an understanding of the appropriate level of contexting, since the acquisition of language and the learning of acts dependent on speech are developed simultaneously. In other words, as children learn to talk, they also learn the fundamentals of social action and interaction. How much or how little contexting is necessary for understanding varies from culture to culture.

Face-saving: This is the value people attach to the maintenance of status and respect and is closely allied to contexting.

Working in a global environment, it is necessary to know the level of contexting in a culture, on a continuum from *low* to *high.* In a low-context culture such as the United States, people rely on verbal self-disclosure to communicate their primary messages. This approach is direct. In high-context cultures people use other means besides verbal disclosure to communicate the message. This approach is indirect. Directness in low-context cultures is a virtue; indirect communication often tries the patience of those involved, and can be interpreted as a waste of time. Indirectness is of course a virtue in high-context cultures, and directness is often interpreted as uncivilized or rude.

As an American manager learning business communication, your course or seminar and the business communication text presented the value of the direct approach. Most likely it extolled the virtue of the Inverted Pyramid as an effective model of business communication to place first things first. You might be surprised and embarassed at the result, then, if you take the direct approach in a high-context culture only to find that the person across the table interprets such directness as very offensive and rude.

Edward T. Hall (*The Silent Language* Doubleday, 1959; *The Hidden Dimension* Doubleday, 1966; *Beyond Culture* Doubleday, 1976; *The Dance of Life* Anchor/Doubleday, 1983) coined the term contexting. He used a diagram to show the relationship between information, context, and meaning. Figure 3.1 shows the relationship of context and the amount of

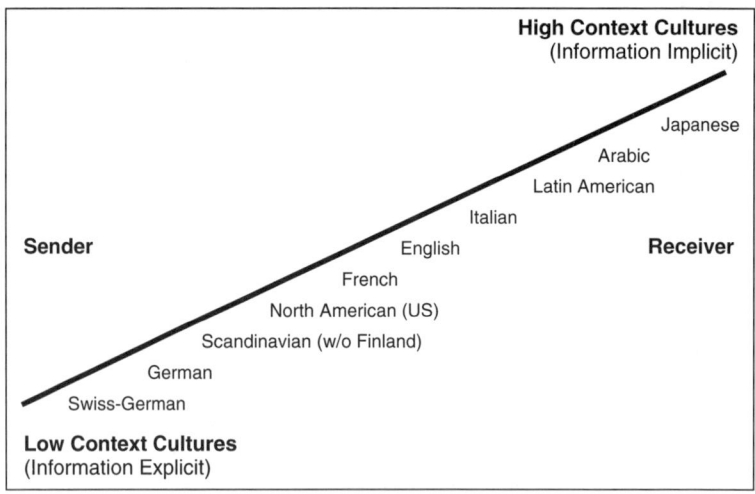

FIGURE 3.1 Edward Hall's Combined Context and Information Triangles and Communicated Meaning Square Indicates that the Higher the Context Culture the Greater is the Shared Meaning

Source: Edward T. Hall

information people store or transmit to communicate meaning.

According to Hall, contexting combines and balances the amount of information stored or assumed, and the amount of information transmitted or shared. Communication in cultures can be expressed along a continuum from a high-context communication in which a great deal of information is shared, to a low-context communication in which the people involved share almost no information. In other words, as context is lost, information must be added to maintain the same level of meaning. In short, meaning cannot exist without both context and information being at appropriate and balanced levels.

Applying the concept of context to international cultures is central to understanding the cultures you encounter in doing business globally. Just as we have mentioned the importance of language and social organization, so too the differences in context and face-saving in developing a strategy for international communication are important. Figure 3.2 offers a continuum of the low-context German culture in which information is expressed explicitly, to the high-context Japanese cul-

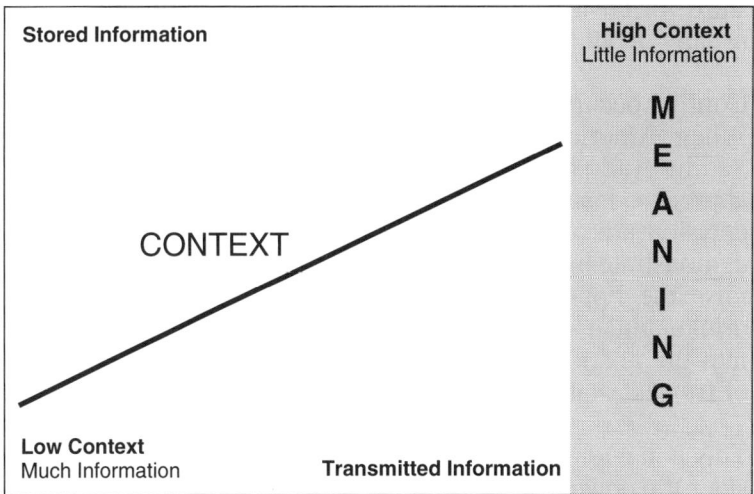

FIGURE 3.2 Ranking of the Context of Cultures Helps in
Understanding the Type of Communication
Appropriate for the Situation

Source: Edward T. Hall

ture in which information is implicit.

As with all the categories and descriptions we are using, keep in mind that individuals within these cultures are likely to be higher or lower than others in the same culture. Communication, particularly across cultures, is very much an art, a learned behavior, and an applied activity constantly changing and evolving.

How does contexting influence behavior? Let's look at how the concept can have an impact on your personal relationships with others, the way you think about the spoken word, the way you believe in the law and contracts.

Most cultures — I'm hesitant to say all — place a premium on a strong personal relationship before they enter into a business arrangement. This is particularly true of high-context cultures because the people in them depend a great deal on shared information, which in most cases becomes known as they develop a strong personal relationship. Low-context cultures such as the United States depend less on these relationships for business because the information communicated is usually detailed and explicit.

Conflict occurs when low- and high-context cultures interact. When Americans do business with Koreans, Thais, and Saudis (all high-context cultures), their brash, direct approach, insistence on data, and need to come to a quick decision are often seen as insincere. Americans who fail to demonstrate interest in the person from a high-context culture miss the opportunity afforded by the practice of sharing drinks, lunch, or a round of golf. Business can be integrated into the relationship only after a strong personal foundation of trust is established.

This behavior carries over to the way different cultures consider the power of the written word, particularly in contracts, codes of behavior, and the law. Americans take great pride in the notion that the United States is a government of laws, not men. This, of course, is the low-context pole of a culture's behavior. The law, or any set of regulations, rules, or corporate policies in low-context cultures governs the interactions of individual people in those societies.

To illustrate, a point of contrast is British versus US law. You might think on the face of it that the two approaches to a similar incident would also be the same since the British and the Americans share so much — language, religion, art, literature, history. But the British are high context and Americans are low. For example, if you owned a business and the walkway outside was covered with ice, in the most states in the United States it is your responsibility to make the walk safe. In the high context of British culture, the person walking has the ultimate responsibility to be careful on the ice, or to stay home. In the United States, the company is slapped with a lawsuit. In England the person goes home to fetch an ice pack for the bump, and maybe a powder for the pain; recourse to

BEHAVIOR TOWARD THE LAW AND SPOKEN COMMUNICATION	IN HIGH-CONTEXT CULTURES	IN LOW-CONTEXT CULTURES
Emphasis on written word	Weak	Strong
Compliance with rules, laws	Loose	Absolute
Contractual agreements	Loosely binding	Binding
Personal promises	Binding	Not binding
Use of the spoken word	Low	High
Use of nonverbal communication	High	Low
Silence as communication	Valued; essential	Unproductive; uncomfortable
Use of detailed data	Low	High
Interpretation of message	Subjective; low literal interpretation	Objective; highly literal interpretation
Communication strategy	Indirect; implicit	Direct; explicit

FIGURE 3.3 Behavior toward the Law, the Written Word, and Verbal Communication in High- and Low-Context Cultures

the courts being unthinkable. Figure 3.3 contrasts attitudes toward the law in low- and high-context cultures.

The concept of contexting also applies to the way people in diverse cultures use and view the written and spoken word. Low-context cultures such as Germany place a premium on rules, regulations, policies. It is the Germans in the European Union who have driven such regulations as ISO 9000, the international rules to which any business must comply to be able to operate in the countries of the European Union.

Take a simple business lunch between an American and individual representatives of high-context cultures such as Japanese, Italian, French, and Kuwaiti. To illustrate the point, the Japanese would use silence and controlled listening, often agreeing with a yes, not to indicate agreement, but rather that he is paying attention to what is being said. Americans often use a form of this themselves in subvocalisms such a uh, and uh-huh often accompanied by a nod of the head. The art of silence, or extraverbal communication, is called *haragei,* or "belly art." The belly is the equivalent for the Japanese of the *heart* for Americans, the portion of the body that houses the person's essence. The art is the intuitive understanding and tolerance gained through experience. Americans might call it wisdom.

Lunch with the Italian or Frenchman might have a great deal of animated conversation and gestures. However, the conversation will be general and rhetorical, with little or no emphasis on detail or practicalities. They are paying attention to the form of the communication, in much the same way they might write a business letter full of formality and polite, accepted phrases with information presented indirectly.

With the Arab, the conversation might be full of fantastic exaggerations and metaphors, leaping from one topic to another in what some have described as loops of thought that move away from the topic and return again more poetically than logically. This is another way of building a relationship indirectly from a high context culture perspective.

SAVING FACE — AN ESSENTIAL PART OF CONTEXTING

People all over the world seek to preserve their outward dignity or prestige — face-saving. Cultures, however, vary in the emphasis or importance they place on face-saving. As a concept, it is directly related to high- and low-context cultures. Figure 3.4 shows the direct corelation between high-context and high face–saving cultures.

High face-saving cultures have these characteristics:
- high contexting
- an indirect strategy for business communication
- tolerates a high degree of generality, ambiguity, and vagueness
- considers indirect communication to be polite, civil, honest, and considerate
- considers direct communication offensive, uncivilized, inconsiderate
- uses few words to disclose personal information

Low face-saving cultures have these characteristics:
- low contexting

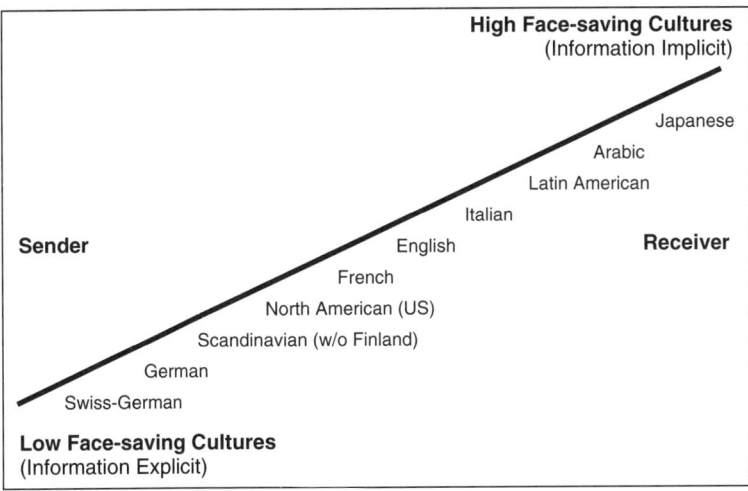

FIGURE 3.4 High Face-Saving Cultures are Directly Related to High-Context Cultures

- a direct strategy for business communication; con-
  frontational
- very low tolerance for generality, ambiguity, and
  vagueness
- considers indirect communication to be impolite,
  unproductive, dishonest, and inconsiderate
- considers direct communication professional,
  honest,considerate
- uses written and spoken words to disclose personal
  information

Saving face is also allied with the concepts of guilt and shame. Shame is associated with high-context cultures; guilt with low. This makes sense when you consider that low-context cultures value rules and the law; breaking the law or a rule implies a transgression — *sin-and-guilt* — as a mechanism for control. High-context cultures use *shame* as the agent of controlling behavior through face, honor, dignity, and obligation.

APPLYING CONTEXTING AND FACE-SAVING FOR BUSINESS

Numerous engineering situations call for your need to apply your understanding of contexting and face-saving. For example, any technical briefing or presentation can mention problems or failures. Here allowing someone to save face is also part of the solution to the technical problem. Selling, meetings, and negotiations offer similar situations in which engineers can reach a successful technical outcome with the aid of interpersonal understanding of contexting.

The analysis of audience as part of the communication process is essential to meet the challenge of managing effectively in a global environment. First, use your ability to listen and observe to determine if the culture is higher or lower context than your own. Also, pay special attention to the work done by an individual in the culture. An engineer from England, for example, has a great deal in common with a counterpart from India. So conversations about engineering projects may run very smoothly, since the profession the two share acts as a high-context situation, even though they come from different nations. Once you have determined that the context of the culture is either higher or lower than your own, follow the principles presented in Figure 3.5.

CONTEXT RELATIVE TO YOURS	PRINCIPLE	COMMENT
Lower	Direct Approach	Less flexibility demands that you express a position and announce your plans for the discussion
	Rude Behavior	People from a lower context culture may seem rude or blunt to you; argument and persuasion are O.K.; attacks on ideas are not personal
	Importance of Rules	Adherence to rules is more rigid than your own
	Explicit Communication	Assume that everything must be put into words to be understood
	Separation of Business and Personal Relationships	They tend to empasize rules over social and personal relationships
Higher	Indirect Approach	Patience is a virtue since business cannot go forward without establishing a firm relationship
	Importance of Face-saving	Pay utmost attention to how your message may be received because every comment will be noticed
	Flexibility of Rules	Make sure you know the difference between the outward appearance and the truth
	Implicit Communication	Form is almost more important than content; pay attention to implications and inferences
	Personal Relationships Essential to Business	Build a personal relationship of trust and shared interests

FIGURE 3.5 Concept of Context Applied to the Business Situation

Understanding and applying the concept of contexting and face-saving will become more and more important as engineering projects reach into the emerging markets and future economic giants of the next century — Africa, China, India, Eastern Europe, Latin America.

CHAPTER 4
Understanding the Impact of Technology and the Environment on Working Transnationally

Most of us take for granted the environment where we live and work. After all, it is what we arc accustomed to, almost unnoticed because it is so familar and unobtrusive that we put it on like an old pair of comfortable shoes. The environment of work can be thought of as having at least three dimensions:
- technology, or the things we make
- nature itself, or the physical area we and the business are in
- the social or civil setting we are in

Also, in working transnationally, keep in mind that not only will you encounter differences in the environment, but in how people throughout the world perceive the environment itself.

The Technical Environment. Americans, particularly American engineers and scientists, take for granted that business runs on technology. You have had a personal computer since the early eighties; a centrally located fax machine at work since the fifties; dependable telephone service, access to electonic databases and to the Internet for a decade or longer if your corporation was involved with government contracts; well-lit offices, dependable transportation to and from work; a social and educational infrastructure that provided you and your employees with stable access to food, clothing, and shelter. You have come to count on and often take for granted all of these things and more, the thousands of items that support and are integral to the conduct of business in the information age.

When you do business globally, get set for sporadic electricity; undependable phone service; scarce sources to repair computers, fax machines, photocopy machines; unreliable transportation; a deteriorating urban infrastructure; primitive conditions in rural areas.

The man-made environment of business around the world may not be as sophisticated or as dependable as the work environment you have left behind. Be prepared for even the smallest task such as copying a document to take longer than you expected. In some countries in the Middle East, the government regulates technology such as access to copy machines. They put a tax on each copy. Even in England the cost per copy is many times that of the cost in the United States. Sending a fax, which most Americans see as a cost-saver, may be more expensive in some countries because of regulations, limited access, telephone charges, and taxes. Access to computer networks throughout the world may also be regulated by the central government as a means of control or censorship. It is best to check on such uses of technology and the regulation of machines and access to information before you go. Murphy's Law holds when it comes to using technology in an unfamiliar place.

Allow me to give you a personal example. I was asked to lecture in Moscow in the summer of 1992 by the Popov Society, the sister to the IEEE/PCS. As one of the places I would speak, they suggested that I come to the Russian Academy of National Economy, the Russian equivalent of MIT's Sloan School, and discuss a 1993 International Conference on Professional Communication in Philadelphia that I was directing and planning. I had prepared a videotape that I had used in the US to promote the conference and to introduce the highlights of the city. It was a standard VHS format, and of course I asked if they had the capability to play it, because I knew that in Europe, TV uses the PAL system. My hosts reassured me that the videotape format was no problem. I wanted to try the tape, but they insisted that it was not necessary — the lunch, the toasts, the meeting with Prof. Anatoly Modin, Director of the center, was much more important.

My lecture began and I signaled for the tape. Nothing but snow and waves on the screen. It was, as I had suspected, a PAL system. They said no problem, disappeared with the tape and suggested that I continue. Minutes later they reappeared with two tapes, and popped the new one in the machine. It played, but without the color. They were able to make the translation to PAL almost instantly. Remember, however, most organizations the world over will be lucky to have a good slide projector or overhead projector for your presentation. Always assume that the technology will not work and plan accordingly if you are making an important presentation. In my case, contexting and face-saving and relationship building were much more important than the content of my presentation and tape.

Plugs, electric cycles, voltages — check your equipment before you go. One of the real benefits of a world marketplace for the global traveler is that many appliances and office products are made to work in an environment of diverse technologies. Appliances sold for the European Union market come without electrical plugs so the buyer can supply one that fits the outlet they have at home.

The issue of workplace safety will certainly come up for engineers. The rules and regulations that American business follows are some of the most demanding in the world. They were developed as a result of owners and employees and the general population placing a high premium on health and safety at work. Other cultures do not have such values for the well-being of the individual. As a consequence, their rules and safety laws may result in manufacturing facilities that seem like something out of nineteenth century sweatshops or from a Dickens novel set in England's industrial revolution.

The tools and technologies may also be adapted to the physical environment because of the access to natural resources or the way the people perceive the role of technology in their culture. Expect that you will encounter significant differences in the way buildings are constructed; the way offices are lit; the dependability of electric, telephone, and other services;

the quality and type of furniture; the actual size of the rooms and their configuration; windows or the lack of them; security measures or the lack of them.

The cultural attitude toward technology is particularly important for engineers to understand when managing globally. There are three ways that cultures seem to view technology: they seek to *control their environment*; they see *control as fruitless or undesirable*; or they seek *harmony with their environment.*

Control cultures such as Canada, the United States, Germany, Great Britain, Australia, the Benelux countries, Scandinavian countries, and Israel see technology as a means to realize their belief systems of self-determination. Technology is a means to achieve cultural goals. They use logic and analytical thought to explain the world. Technology through research and development becomes the way to shrink distances, cure disease, hold back the tides. For members of control cultures, other cultures are less advanced and less civilized, and industrialization is considered a mark of accomplishment.

Cultures that find control of the environment impossible or undesirable, see technology as a waste of time because they believe that what men do will have little impact on the outcome. From their perspective, *technology is at best neutral, at worst harmful or negative.* Any anthropologist will report that in such cultures, technology changes the culture itself and in some cases can destroy the culture. The change can be the source of resentment between two cultures as a result of the technology transfer. In such countries, technological changes may be perceived as a threat to the authority of their god or religion. Superstitions are also part of these cultures, and need to be understood rather than dismissed as backward and uncivilized. And thirdly, cultures such as the Japanese, Chinese, and Indian see themselves in *harmony with their world,* neither controlling it nor controlled by it.

The Physical Environment. The physical elements of the land and climate of a country play an important role in the way an area of the world develops. Climate and physical environment have an impact on the way people think of themselves, and thus affects their behavior. We have all heard cliches, that people from countries closer to the equator are slower and less productive than their counterparts who live farther north.

Americans, unlike other nationalities, do not give much consideration to climate. They use technology to control the temperature in their cars, their homes, and their offices. They dam rivers, drain swamps, put up levees to hold back seasonal flooding, and irrigate arid farmland. In most of Africa, South America, Eastern Europe, and Asia, nations and their businesses are at the mercy of their climate. Climate becomes a critical factor in how they do business. Business adapts to the environment. Such an attitude toward the elements is necessary for understanding local business practices.

The natural features of an area are just as important to understand how business is conducted. Mountains, rivers, bodies of water, and distances within a country all play a part in determining the way people in that area think. Even a seemingly homogenous culture as the Swiss has influences from the French, Italian, and German largely determined by mountain ranges. The same factors determine the provincial difference in France, Spain, Germany, and The Netherlands.

Physical distances are also something to consider. Americans think nothing of commuting fifty or more miles a day in the open parts of the midwest and the west. For an Frenchman, Italian, or Englishman, the idea of such a journey at all, much less every day would seem burdensome, particularly because many of them have not traveled that far from home in their lives.

The existence of a nation's natural resources or lack of them also has an impact on the culture and shapes the culture's relationship to these resources. The forests of Rocky Mountains of Canada and the United States, for instance, allow people there to think of wood as plentiful enough to burn for a fuel to heat homes and as a raw material to be exported.

The Social Environment. The makeup of a society, apart from the issues we have already discussed, also has an impact on doing business globally. Issues such as population density and size often determine whether or not a country is an export or import nation. Alliances such as the European Union (See Chapter 12 below in Section III) are essential for doing business. In that case alone, the countries have agreements that set forth mandatory standards for people to be able to do business in those countries.

Population density also has an impact on the use of space. Cities such as New York and Tokyo place a high premium on space, since space is so scarce. In the United States, England, and Japan populations are not evenly distributed throughout the country. Territory outside city limits can be sparsely populated bordering on being relatively uninhabited, while urban dwellers seem to be literally stacked on top of one another or packed like sardines in subway cars and commuter trains.

CHAPTER 5
Understanding Concepts of Power, Influence, and Authority in a Global Environment

Understanding the workings of power, authority, and influence is another concept that needs to be mastered for managing and communicating successfully in a global marketplace. Power in western countries is related to the ability to make a decision, and the ability to implement it. Power is often seen as separate from the person. It is something that goes with the title or the office. Asian cultures characterize power and the decision-making that is so desirable in the West as something to be avoided. The Asian practice of consensus decision-making reflects the desire to be part of the group rather than the leader.

The recognition of power is reflected in the language of a country. The use of titles is a way many cultures recognize status. Of course, the use of titles and the meanings of their use vary from society to society.

Americans, who are raised in a democracy, can approach this topic by discussing the ways different cultures and groups handle equality or inequality. Common sense tells you that in any group of people some are taller, stronger, or smarter. Some have more power; some have more wealth; some have more status; and some have more respect. The combination of status, power, and wealth can create problems in some societies. For Americans, success in one area does not mean that success will come in other areas. Someone can be successful in one area, but not necessarily in all. A politician who uses his or her position and status to gain personal wealth is seen as violating society's concept of equality. Societies that have a large middle class, a group between the "haves" and the "have nots," also have a high value for equality.

For other societies, the distinction between the haves and the have nots is stark, and the notion of equality is superseded by the acceptance of inequality. How can you determine this factor in traveling and working transnationally? Geerte Hofstede's benchmark study of IBM employees in fifty countries and three regions (*Cultures and Organizations,* 1994) used three questions to determine a notion he called the **power distance index.** The survey questions were as follows:

- How frequently, in your experience, does the following problem occur: employees being afraid to express disagreement with their managers? (mean score on a 1-5 scale from "very frequently" to "very seldom").

- Subordinates' perception of their boss's actual decision-making style (percentage choosing either the description of an autocratic or of a paternalistic style, out of four possible styles and a "none of these" option).

- Subordinates' preference for their boss's decision-making style (percentage preferring an autocratic or a paternalistic style, or, on the contrary, a style based on majority vote, but not a consultive style)

Hofstede's scores led to a ranking of countries that shows the relative positions of each in relation to one another. Figure 5.1 lists a selection from the list of 50 countries and three regions.

The degree of inequality or equality in a country allows us to see the "dependence" relationship between bosses and subordinates. So in small power distance countries such as Sweden or Germany, therefore, the dependence is limited. People in those countries relate to one another readily and do not have a great fear of contradicting a superior. Professionals in small power distance countries perfer relationships that are consultive or interdependent.

SCORE/RANK	COUNTRY or REGION	POWER DISTANCE INDEX SCORE
1	Malaysia	104
2/3	Guatemala and Panama	95
4	Philippines	94
5/6	Mexico and Venezuela	81
7	Arab Countries	80
8/9	Equador and Indonesia	78
10/11	India and West Africa	77
12	Yugoslavia (former)	76
14	Brazil	69
15/16	France and Hong Kong	68
17	Columbia	67
20	Belgium	65
24/25	Chile and Portugal	63
26	Uruguay	61
27/28	Greece and South Korea	60
31	Spain	57
33	Japan	54
34	Italy	50
35/36	Argentina and South Africa	49
38	USA	40
39	Canada	39
40	The Netherlands	38
42/44	Germany and Great Britain	35
47/48	Norway and Sweden	31
52	Israel	13

FIGURE 5.1 Hofstede's Power Distance Index Shows a Relative Degree of Inequality in a Country or Region Compared with Other Countries

Source: Hofstede

Countries with high power distance scores such as Arab countries and France show a high degree of dependence. If they reject the dependence it is counterdependence, or dependence with a negative sign.

In practical terms, what would happen if you were to take some popular American management techniques such as TQM, re-engineering, empowerment, or team-centered management to another country? These collaborative approaches and techniques achieve positive results in an environment of high equality and in a society that values independence. It should come as no surprise to you and your company, then, if the programs are an utter failure in relatively high power distance countries such as Mexico, France, or India.

The concept of power distance or inequality also occurs within countries according to social class, education level, and occupation. Therefore, many American engineers who are college graduates and often hold graduate degrees, but consider themselves part of the American middle class, are startled to be treated as part of a higher social class in higher power distance countries.

The manifestation of inequality at work is a combination of numerous forces in each society — education, family life, and occupation. If we keep in mind that generalities about a society can aid our understanding of them rather than present an accurate description of the society itself, then we can make generalizations about cultures that have high and low power distances. Figure 5.2 shows some contrasts between small and large power distance societies.

The concept of power distance translates into a management or communication styles popularly known as Theory X and Theory Y and Theory Z. Think of heirarchical cultures as generally having large power distance. These cultures are Theory X and reflect old-style authoritarian leadership and communication behaviors shown in the LARGE POWER DISTANCE column of Figure 5.2.

A contemporary management model that values a team-orientation and collaborative effort is relected a THEORY Z management style. The SMALL POWER DISTANCE column of Figure 5.2 lists the characteristics of such leaders.

Expect conflict if you practice managing and communicating on democratic or small power distance principles, and the people in the country you are in are used to a large power distance between bosses and subordinates. Avoid conflict by understanding the concept of authority and how it is manifested in daily behavior through power distance.

GROUP OR CONCEPT	SMALL POWER DISTANCE	LARGE POWER DISTANCE
Inequality	Effort to minimize	Expected and desired
Dependence	Interdependence among less and more powerful people	Less powerful people polarized between dependence and counterdependence
Parents	Treat children as equals	Teach children obedience
Children	Treat parents as equals	Treat parents with respect
Teachers	—Experts who transfer impersonal truths —Expect initiatives from students in class	—Gurus who transfer personal wisdom —Expected to take all initiatives in class
Education	More educated people hold less authoritarian values than less educated persons	Both more and less educated persons show almost equally authoritarian values
Hierarchy in organizations	Means inequality of roles established for convenience	Reflects the existential inequality between higher-ups and lower-downs
Centralization	Decentralization is popular	Centralization is popular
Salary range	Narrow between top and bottom of the organization	Wide between top and bottom of the organization
Subordinates	Expect to be consulted	Expect to be told what to do
Ideal boss	A resourceful democrat	A benevolent autocrat or good father
Privileges and status symbols	Frowned upon	For managers both expected and popular

Figure 5.2 Differences Between Small and Large Power Distance Cultures Helps to Understand Personal Interactions at Work

Source: Hofstede

CHAPTER 6
Body Language and Non-verbal Communication for Global Business

Discussions in the five previous chapters of these issues:

- language
- social organization
- contexting
- relationship to technology and environment
- power, authority, and influence

and the differings concepts of time which we will discuss in the next section, should be of help in interpreting the nonverbal communication you encounter while working in an international environment. These concepts should provide a context or background for you to interpret the gestures, sounds, eye movements, and facial expressions that you will see.

When you begin your journey overseas you will read books, guides, and tips for behavior in doing business in another culture. Of course that's what brought you to this book in the first place — a desire to understand the ways to behave effectively when doing business around the world. Figure 6.1 offers some examples of typical dos and taboos.

A facial expression as nonverbal communication can have similarities across cultures: a smile, a nod. But the culture has an impact on how much or how little emotion is displayed. Some international culture experts have observed that the development of nonverbal behavior is linked to the culture's historical development in a way that is similar to the etymology — the semantic derivation and evolution — of the language spoken there. Some also observe that a person's body language is an analog for that person's social system. So the amount of information communicated nonverbally may be up to and more than 50% of the message.

GESTURE	COUNTRY	DO or TABOO	COMMENT
FACE Wink	Australia	Improper to wink at a woman	Facial expressions and eye movement and contact may have several expressions that are universal — smiles, sadness, anger, curiosity. How each is interpreted in the specific culture is a matter of learned behavior. Patient observation is your best approach to coping with the differences and similarities.
Blink	Taiwan	Considered impolite	
Ear pull	India	Sign of sincerity	
Ear pull	Brazil	Signifies appreciation	
Nose tap	Britain	Secrecy or confidentiality	
Nose tap	Italy	Friendly warning	
Kiss of fingertips	Europe (throughout)	"Aah, BEAUTIFUL!"	
Head nod	Bulgaria & Greece	Signifies "No"; "Yes" in most other places	
Head tap	Argentina & Peru	"Think" or "I'm thinking"; "He's crazy" in most other places	
HAND/ARM V sign (out)	Europe	Means "victory"	As we have discussed above, the concept of contexting can help you to interpret hand gestures and other body movements. The power relationships and the social organization can also be manifested in the
V (palm in)		Roughly, "Shove it"	
Beckon	Middle & Far East	Insulting to call someone using a finger or fingers	
"OK" sign	Brazil & Germany	An obscene gesture	
	Greece & Russia	Impolite	

GESTURE	COUNTRY	DO or TABOO	COMMENT
Pointing	Middle & Far East	Impolite	gestures and nonverbal communications of a group of people. How much or how little emotion is displayed in a person's face is a mix of individual personality, as well as the influence of the culture the person was brought up in.
Third-finger salute	Universal	The Romans called the third finger the "impudent" finger.	
Thumbs up	Australia	Rude in Australia; "OK" in most other places	
Waving	Greece & Nigeria	A serious insult	
	Europe	"Good-bye"; can mean "no" waving the whole head back and forth	
BODY Bowing	Japan	Shows relationship of respect and status; depth, duration, and number of bows — lower, longer, more to show respect for superiors in age, power, experience	Touching and body language is also related to the way a particular culture interprets the notion of privacy and personal space. How close or far people can be in a business setting is determined by the culture.
Shoe soles	Middle East & Asia	Sign of disrespect to show the soles of your shoes; keep feet flat on the floor	
Hugging	Latin & Slavic countries	The equivalent of a handshake	

FIGURE 6.1 Some Body-language "Do's and Taboos"
Source: Axtell, Do's and Taboos Around the World

Thus, the chances for miscommunication and misinterpretation are great and increase when the communication is in a multi- or cross-cultural environment.

Nonverbal communication is difficult enough to understand in your own culture for the simple reason that such behavior is learned on a daily basis from birth. Nonverbal communication is almost like walking or talking — we take its structure and impact for granted because it is second nature to us. Consider the following to help in your understanding of others:

- Kinesics, or movement
- Appearance
- Eye movement, or oculesics
- Touching behavior, or haptics
- Space usage, or proxemics
- Non-word sounds, or paralanguage
- Interpretation of color, smell, symbols, numbers

Kinesics, or movement. Movement in any context is affected by an individual's personality, the situation itself, differences in gender, and the force of the person's culture.

An individual's personality is so specific that trying to make a generalization would be a waste of energy. You can, however, note that a person you do business with behaves in a particular and unique way.

Situations have an impact on how we behave... meetings in the office, speeches before a large audience, face-to-face meetings over lunch, chance meetings in the hallway of the company. Gestures in small meetings may be understated compared to the sweeping, dramatic movements to make a similar point when motivating a large gathering.

Men and women also use movement differently. Some attribute women's behavior in the workplace as different because of the subordinate position of women in many cultures. We hope this is changing as a generation of American female managers and executives redefines the behavior of working women throughout the world.

Culture also determines the meaning of movement. Whether a person moves rapidly, often, or in an animated way is a learned cultural behavior. The cultural differences may be easier to observe and interpret than personality, situation, and gender.

Kinesics can be thought of as serving several purposes in communicating a message:

- replacement of words with a direct verbal equivalent, or an emblem
- movements that reinforce a verbal message, or illustrators
- facial expressions, or affect displays
- listening signals, or regulators

Figure 6.2 provides some examples of kinesics.

David Victor makes a profound observation about movements, but he also offers a strategy for operating successfully in other cultures:

MOVEMENT	EXAMPLE
Emblems	Sign to signify "OK"
Illustrators	Gesture to accompany "come here" or "stop"
Affect	Facial expressions that display: anger, fear, happiness, sadness, surprise, disgust, contempt, interest, bewilderment, determination
Regulators	Nodding of the head for Americans signifies assent, for Japanese it merely says "I am listening"; usually head nodding in most cultures is accompanied by eye contact

FIGURE 6.2 Identifying the Purposes of Nonverbal Communication Helps You Understand the Behavior You Observe in Different Cultures

The key point is *not* to memorize a list of customs and taboos around the world. Such a list would reduce cross-cultural nonverbal communication to either a bewildering array of unintelligible practices or a collection of quaint "do's and dont's." Instead, international business communicators should be aware of different kinesic emblems in order to recognize them as they occur, expanding knowledge of them as one does the words of a language. (Victor, 192)

Appearance. Within a culture and across cultures, the way a person looks — his or her appearance — makes a strong nonverbal statement. Americans, as well as most other cultures, place a great deal of emphasis on looking professional at work. The way a person looks is a function of the way they dress and their physical appearance.

Physical appearance includes skin color, eye color and shape, type and color of hair, and body size and structure. People who work in most global organizations generally do not accept the cultural bias that results from stereotyping based on physical differences. This does not mean that they ignore racial prejudice in a country; it suggests that they become aware that in other countries the value placed on race and other factors is different from their own.

Dress can be a way to communicate as well. A uniform worn by soldiers, UPS employees, or McDonalds workers communicates the power, rank, and authority within the company. Uniforms sidestep physical differences, and because they are the same, can neutralize power and difference.

Eye Movement, or Oculesics. From the first moments of life communication is established nonverbally with the eyes. Babies interact with their mothers during breast feeding, communicating with their eyes.

The most common eye behavior is eye contact. People perceive it as either too direct or too indirect according to their own cultural expectations, rather than that of others.

Americans look each other in the eye when speaking. Such behavior indicates active listening to another American. However, direct eye contact is perceived as rude by the Japanese. Like the Japanese, many cultures link eye contact with age differences and relative status. People of perceived lower status look below the gaze of the other.

Gender also plays a role in eye behavior. American women are often offended when in France or Italy because of the way men look them over for such a long time, and so thoroughly. American men only look glancingly at women.

Touching Behavior, or Haptics. In the workplace a hand-shake, a pat on the back, a hug, and a kiss are examples of touching behavior, or haptics. In general hostile touching — punching or kicking — is rarely associated with normal workplace behavior. Touching behavior exhibits progressive intimacy from functional/professional to social/polite to friendship/warmth to love/intimacy to sexual arousal. The first three are appropriate for the workplace. The culture of an individual helps that person determine the difference between the social and the friendly.

Space Usage, or Proxemics. People throughout the world structure the space around them differently. The size of an office, the furniture, and the doors and windows are defined by the culture. The notion of walls and privacy are also part of the culture. The contemporary office in most major American corporations has few floor-to-ceiling walls; instead it is a sort of cubical that affords privacy as one sits and works. The arrangement is a balance between an individual's right to privacy and the membership in the company culture.

In interpersonal communication, the physical space between speakers — the proxemics — adds to the message. The culture again determines whether the distance is too close or too far away. Distance can be thought of as showing degrees of intimacy, as depicted in Figure 6.3.

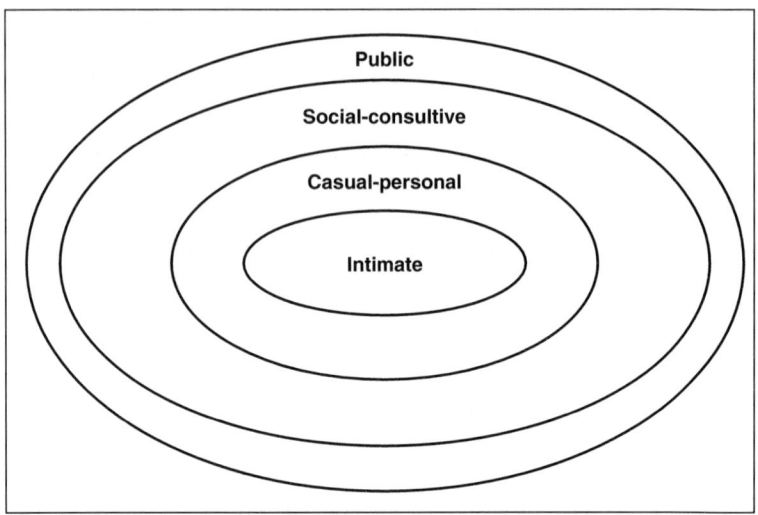

FIGURE 6.3 The Boundaries Between These Spaces Is
Different From Culture to Culture

The intimate space is reserved for the people closest to you. In general, this space is extremely inappropriate in the workplace. Friends and relatives are usually comfortable in this range. The workplace is generally characterized by the social-consultive distance. The public space at work is limited to formal presentations. The actual distance for each of these categories changes according to the culture, and can be the cause of some discomfort when meeting and talking with people from different cultures.

To illustrate the point, imagine a conversation between a North American and a Latin American. The North American is generally comfortable in the social-consultive distance of "arm's length" in conversations. The Latin American, as well as a southern European and a Middle Easterner, is comfortable at a much closer distance for that category. The North American, to ease the discomfort, will take a step back to make the distance greater. Sensing a bit of coldness or rejection, the Latin will step forward. The dance will continue until the North American is literally backed against the wall, or in an extreme case, puts his hands up to push the other away.

Non-word Sounds, or Paralanguage. People also communicate with sounds that are not words. These vocalizations are generally universal, such as giggling, yelling, moaning, whining, crying, and laughing. Pacing sounds such as *uh* or *um* are usually considered vocalizations.

The way a person speaks is also a factor in nonverbal communication. The pitch, the pace, the articulation, the rhythm, and the volume are often culturally determined. Sometimes these factors change within cultures as with accent. Southern Germans speak more slowly and with different sounds for the same words as Germans from the north. A deep, breathy voice for an American female is not always sexy, but it is usually interpreted that way.

However, people can usually determine, based on the way he or she speaks, the speaker's emotional state, honesty, status, and attitude.

Interpretation of color, smell, symbols, numbers. The way people interpret colors, smells, symbols, and numbers differs from culture to culture.

White is a symbol of mourning in Japan, quite the opposite of the purity and innocence it stands for in the West. A Japanese automaker tried to understand why its cars painted bright orange with blue interiors — a smash at home — were a flop in the United States. The next year it used "American" colors.

Smell, or rather the absence of smell, is the message in the Americas, northern Europe, Japan, and Australia. Bad breath and body odor are offensive in these cultures. But in many Middle Eastern and African countries natural odors are part of communicating many emotions, from fear to friendliness. People from these countries might even consider the use of a deodorant or mouthwash an effort to cover up true feelings.

Symbols represent something in the culture. These may be the flag of the country itself, a cross or star or crescent for religion, or the image of a national hero. Such nonverbal communications are very powerful for those who value the symbol and believe in what it represents. Other symbols, such as traffic signs, are created to be understood throughout the world.

To become familiar with the nonverbal communication patterns of the country you plan to visit, a useful strategy is to get a copy of the magazines and business newspapers from that country and observe the way people look in the pictures and the way they are depicted in the advertising. While in Moscow in 1992, shortly after the revolution that ousted Gorbachev and put Yeltsin in power, I saw a commercial for a GM Chevy. The actors were dressed like gangsters from Chicago circa 1929. I saw the same gangster motif at the Moscow Circus. The message: capitalism and wealth are somehow linked with organized crime.

The hidden messages of nonverbal communication present another challenge as you live and work overseas. When you are in a situation that requires you to be too close to someone from another culture, or to greet someone with a hug and a kiss on the cheek, the best way to deal with the discomfort you may experience is to be a bit flexible and accommodating going into those countries in the first place. Your ability to delay judgment may be very useful for you to cope with differences you encounter in the culture. (For a related discussion, see Chapter 11 on Dealing with Culture Shock.)

CHAPTER 7
Concepts of Time and Communication in an International Environment

In the Caribbean, harried tourists just off the plane from New York rush to line up for a taxi. Their body language and tone of voice communicates impatience and anxiety as they ask when the van from the hotel will arrive. The man in charge looks absentmindedly at the water by the dock and says in a lilting voice, "Come soon." The Americans become more anxious and impatient because they have not been given a precise time. And the taxi dispatcher sees no reason to hurry because the van is on its way over some rather treacherous mountain roads and cannot go any faster than it is going.

Business travelers to Moscow almost uniformly report that the meetings, rides, tours, and appointments that are scheduled with such precision when planning a business trip, deteriorate into an ad hoc fire drill as appointments run late, cars do not show up for transportation, or the people you are meeting with continue to talk long after the scheduled time has elapsed. For Russians, such loose adherence to the schedule causes no difficulty.

Such lack of punctuality drives the British, Americans, and Germans up a wall. Why?

Time, as anyone who has heard of Einstein or Stephen Hawking can attest, is relative. Our concepts of the passage of time are culturally defined. Clocks and calendars are the artifacts that civilizations and societies use to help them mark the passing of time. Victor (*International Business Communication:* 228-244) discusses two classifications of business cultures in terms of time: monochronic and poly-

COUNTRIES WITH MONOCHRONIC BUSINESS CULTURES	CHARACTERISTICS
United States, Canada (English-speaking), Great Britain, Australia, New Zealand, British South Africa, Sweden, Norway, Denmark, Iceland, Germany, Luxembourg, Austria, Netherlands, Switzerland (German)	Interpersonal relations subordinate to schedule Schedule coordinates activity; appointment time is rigid One task handled at a time Breaks and personal time are sacred regardless of personal ties Time is inflexible; time is tangible Work time is clearly separable from personal time Activities are isolated from organization as a whole; tasks are measured by output in time (activity per hour or minute)
COUNTRIES WITH POLYCHRONIC BUSINESS CULTURES	**CHARACTERISTICS**
Italy, Portugal, France, Spain, Greece, Brazil, Mexico, Turkey, Egypt, India; generally most of Latin America, Arabic-speaking nations, Africa, southern and western Asia, the Caribbean, southern Europe	Preset schedule is subordinate to interpersonal relations Interpersonal relations coordinate activity; appointment time is flexible Many tasks are handled simultaneously Breaks and personal time subordinate to personal ties Time is flexible; time is fluid Work time is not clearly separable from personal time Activities are integrated into organization as a whole; tasks are measured as part of overall organizational goal

FIGURE 7.1 Characteristics of Monochronic and Polychronic Time Concepts in Business Cultures in Various Countries and Regions

Source: Victor, International Business Communication

chronic. Figure 7.1 presents the characteristics of monochronic and polychronic business cultures.

How can you cope, as an American, when working in countries that have a polychronic business culture? As with all the other topics we have discussed in this chapter, be flexible. Throw out your ethnocentrism and enter the global business community. You will find that when you do, your counterparts in other countries have a great deal more in common with you on a professional level than they do with members of their own cultures and societies who are not in the same profession.

Also keep in mind that leaps in communiction technologies, global computer networks, global news by satellite, and media advances are shrinking the world. Old barriers and concepts dissipate as nations come together.

PART II
WORKING THE GLOBAL ENVIRONMENTS AND CULTURES

- Negotiation and Winning in Global Environments
- Marketing Transnationally
- Managing People Globally
- Dealing with Culture Shock

CHAPTER 8
Negotiation and Winning in
Global Environments

In our contemporary global economy you most certainly face, particularly in engineering, the   prospect of multicultural negotiations. As an American, you might meet with Germans on a Wednesday who use an interpreter and on Friday with a German friend fluent in English with whom you have worked over the years. You may be in a London hotel negotiating face-to-face with Japanese and be served lunch by an Indian waiter who speaks French to the Japanese and English with you. You may have a business meeting with Brazilian engineers at the Dallas/Fort Worth airport. You may have more than a year to negotiate a joint venture with another country or foreign business; you may have only 24 hours to complete a deal by e-mail, fax, telephone, or videoconference. The transnational environment requires you to be ready to negotiate with people who do not share your cultural background.

Negotiation is an art, not a science. The simple reality is that each occasion to negotiate will present unique challenges and can be the cause of great professional and emotional stress. Negotiation implies, for some business professionals, winners and losers. In the contemporary business context, however, an outcome that creates winners on both sides, a win-win solution, is almost universally recognized as the desirable goal of any negotiation.

As difficult as it is in your own culture and in your own language, negotiation takes on added complexity and difficulty in a global and transnational setting. Negotiation is often discussed in terms of a game, with the players seen as adversaries or opponents. Such a win-lose perspective can sometimes prove successful in the short-term in your own culture where you more or less share the same perspective. In the long run, however, that approach may undermine long-term relationships that can result in profitable collaborations on many projects at all levels.

As discussed in the previous section, differences in point of view, and of world view, make the art of international negotiation one of overcoming the differences in objectives and points of view. It is an art that requires skill, understanding, preparation, creativity, and strategy. The process has three main movements: Prepare, Bargain, Agree and Maintain the deal.

PREPARE

The previous sections (Chapters 1-7) of our discussion should provide you with some basic tools for understanding the culture you are dealing with as you prepare for negotiations. Such cultural concepts as uncertainty, avoidance, individualism/collectivism, masculinity/femininity, power distance, short- or long-term perspective are important for you to know as you do your homework. Any communication requires you to know your audience's needs and expectations; international negotiations demand a high level of understanding for the bargaining to be successful.

People from weak uncertainty avoidance cultures may be reluctant to follow procedures consistently, while Americans who have strong uncertainty avoidance cultures might insist too strongly on the rules and regulations, allowing the letter of the law to undermine an agreement.

The classic American/Japanese negotiation is one that illustrates the individual culture versus the collectivist one. And the cultures high in masculinity might attain a paper contract over a low masculinity culture negotiator, but not have a lasting agreement that leads to a productive relationship.

Negotiators from large power distance (hierarchical) cultures may feel little or no need to follow the letter of a contract —

again, leading to a paper contract, but no viable partnership. And people from cultures with a short-term focus, such as the United States, may be frustrated by the lengthy discussions, delays, and tangents that people from a long-term perspective might engage in. Figure 8.1 offers suggestions for getting ready for a global negotiation.

In preparing for a negotiation also consider these five cultural characteristics that have an impact on negotiation:

- the general model
- the role of the individual
- the nature of personal interactions
- the process of interactions
- the outcome.

The General Model. The fundamental concept of negotiations as a process may differ from culture to culture. It may run the range of approaches including some or all of these: distributive bargaining, joint problem-solving, debate, contingency bargaining, and nondirective discussion. The significance of the type of issue may also vary from culture to culture, from the substance of the negotiations to the relationship on which the negotiation is based, to the importance of the procedures, to the personalities involved.

The Role of the Individual. The selection of negotiators usually focuses on people who have knowledge, negotiating experience, the proper personal characteristics for negotiation, and status. The aspirations of the people involved may range from personal to the community. The decision-making process may also run from authoritative to consensus.

The Nature of Personal Interactions. Concepts of time from monochronic to polychronic enter into the dynamics of negotiations. Whether a person has a high or low propensity for risk-taking also applies. And the way people form trust — through shared experiences, intuition, reputation, external sanction — has an impact on how people interact in a negotiation.

BEFORE THE NEGOTIATION	QUESTIONS FOR PLANNING
Make sure that what you are negotiating is negotiable.	Does the problem have a solution? Can differences be reconciled? Does interest in agreement exist? Is this a good cultural fit?
Define what "winning" the negotiation means to you.	Have you considered price, quality, quantity, timing, delivery, warranties, costs, terms of payment, labor arrangements, regulations, "standard business practices" in the other culture? Have your defined what not winning means? Do you accept the notion that no deal is better than a bad deal?
Get the facts.	Have you had a pre-negotiation meeting? Who are the decision makers? How are the decisions made? Budget? Schedule of delivery? What are the company goals and strategic mission? What do you know about the people on the negotiating team?
Have a strategy for each culture and for each phase of the negotiation.	Have you determined how to position your proposal? Is your approach competitive (win-lose) or cooperative (win-win)? What is your opening offer? What concessions do you plan to make?
Send a winning team.	Have you selected the right people for the team? Do they have the authority? Do you plan to send several people? Have you arranged for your own interpreter? Have you planned to leave the lawyers and accountants behind? Have you planned to keep the same negotiator for the entire time?
Allow yourself plenty of time, and more.	Have you multiplied the time you think it will take by three? Have you made reservations to stay for an indefinite period?

FIGURE 8.1 Suggestions for Preparing for Negotiations

Source: Copeland and Griggs, Going International

The Process of Interactions. Culture also determines the way people think about protocol, from informal to formal. Communication can be complex or simple. People from various cultures persuade others through different methods: direct experience, logic, tradition, dogma, emotion, and intuition.

The Outcome. The actual agreement might be implicit or it might be a formal contract. Some cultures, such as Americans, want it all in writing, while others such as the Russians, base the agreements on relationships among decision makers.

BARGAIN

At the beginning of the negotiation, setting the tone is particularly important. For Americans and northern Europeans, the opening is difficult because they are eager to get straight to the point. Resist the temptation to see the initial meetings that might appear social to you as something less than businesslike. These meetings seem informal, but they are opportunities for your hosts to check you out. Be wise and do the same. Be aware that you are at a business meeting and that it is business. Be professional in these situations and personable.

The first phase of introductions is very brief in the United States, but almost everywhere else it takes a long time. Take advantage of the time to do as your opposition does and feel out the players yourself.

When considering the agenda proposed by the other side, consider it carefully. Interpret it. What is in it? What is left out and why? What does it reveal? Study it carefully, and ask for time to consider it if you are concerned.

The physical accommodations can be important in negotiations. Home field advantage can be significant, but you can play on the host's obligations if you are a visitor. Some simple warnings — try not to face the sun in a negotiation, and insist on a quiet room. But be prepared for heat in summer and cold in winter in most parts of the world.

In the negotiation process, the formal process begins with "posturing," general statements used to set a tone and launch the negotiation on the right foot. This is followed by an exchange of information usually through presentations and a round of questions and answers.

The next phase, bargaining, then begins. The result is to forge a deal, each side trying to persuade or manipulate the other. Figure 8.2 offers some advice, as well as some strategies and tactics, for the bargaining phase.

Here are some negotiation approaches you can take with two representative countries in using a culturally responsive strategy. Americans negotiating with the Japanese, according to Stephen Weiss (*Sloan Management Review,* Winter 1994), should:

- Use an introducer for initial contacts.
- Employ an agent the counterpart knows and respects.
- Ensure that the agent/advisor speaks fluent Japanese.
- Be open to social interaction and communicate directly.
- Make an extreme initial proposal, expecting to make concessions later.
- Work efficiently to "get the job done."
- Follow some Japanese protocol (reserved behavior, name cards, gifts).
- Provide a lot of information (by American standards) up front to influence the counterpart's decision making early.
- Slow down your usual time table.
- Make informed interpretations.
- Present positions later in the process, more firmly and more consistently.
- Proceed according to an information-gathering, nemawashi (not exchange) model.
- "Know your stuff" cold.
- Assemble a team (group) for formal negotiations.
- Speak in Japanese.
- Develop personal relationships: respond to obligations within them.

BARGAINING SUGGESTIONS	ACTION STRATEGIES AND TACTICS
Control information.	Answer questions and ask questions diplomatically to get discrete information without tipping your hand. Be careful with proprietary, secret, and confidential information.
Watch your language.	Remember the discussion about words, body language, and meanings. Set agreement on the meanings of words and concepts. Try to use simple language that is free of idioms, metaphors, similes, and other figures of speech that lose in translation or are confusing when translated.
Persuasion is an art.	What convinces Americans does not always convince others. Make an effort to discuss matters on the same level. In making concessions, do so in small and consistent increments. Some American negotiation strategies do not translate well in other cultures, such as the "good guy bad guy" and resolving issues separately in small parts.
Get in stride with the locals.	Other cultures are not linear in thought and decision-making processes. They often see the whole of the problem and consider all discussion pointing to that end. The pace is also different, so take the time to think, like taking a time-out in a basketball game to assess your next move.
Consider the negotiation session as the formal forum and use informal channels.	In many countries, particularly in the Far East, the meeting is the formal and public expression of agreements previously worked out. Come early and stay late to resolve differences.
Give face.	As discussed above, treat your opponent with respect and fair play. Allow everyone to save face.
Deadlock means no winner, but might signal both as losers.	Explore ways to give minor concessions, often through a go-between, that might shake the deal free. Be aware that not every impasse is a deadlock.
Walk away rather than take a bad deal.	Know your walk-away position before you start, and walk away from a bad deal. Be able to say no even when negotiating with cultures that emphasize harmony and good will in communications.

FIGURE 8.2 Suggestions for the Negotiation Process
Source: Copeland and Griggs, Going International

- Do homework on the individual counterpart(s) and circumstances.
- Be attentive and nimble (improvising entails different
  behaviors for different Japanese).
- Invite the counterpart to participate in mutually
  enjoyed activities or interests (e.g., golf).

Americans negotiating with the French, also according to
Weiss, should:

- Employ an agent well-connected in business and gov-
  ernment circles.
- Ensure that the agent/advisor speaks fluent French.
- Be open to social interaction and communicate directly.
- Make an extreme initial proposal, expecting to make
  concessions later.
- Work efficiently to "get the job done."
- Follow some French protocol (greetings and leave-tak-
  ings, formal speech).
- Demonstrate an awareness of French culture and busi-
  ness environment.
- Be consistent between actual and stated goals and
  between attitudes and behavior.
- Defend views vigorously.
- Approach negotiation as a debate involving reasoned
  argument.
- Know the subject of negotiation *and* broad environ-
  mental issues (economic, political, social).
- Make intellectually elegant, persuasive, yet creative
  presentations (logically sound, verbally precise).
- Speak in French.
- Show interest in the counterpart as an individual but
  remain aware of the strictures of social and organizational hierarchies.
- Do homework on the individual counterpart(s) and circumstances.
- Be attentive and nimble (improvising entails different
  behaviors for different French individuals).
- Invite the counterpart to participate in mutually
  enjoyed activities or interests (e.g., dining out, tennis).

AGREE AND MAINTAIN THE DEAL

Before you leave the country make sure that you have a signed agreement. Once you are gone, other matters and more people may get involved. The deal may unravel.

Also make sure that all parties involved agree on the significance of the signed agreement. As we have discussed, not all cultures consider a written contract as binding as Americans do. According to American sales consultant James Kudless, "From a contract management perspective, Swedes try to develop a relationship during the drafting of the agreement. Once the agreement is in place, the terms are never referenced or consulted, and sometimes never recorded. After the agreement is obtained, some Swedish business people believe it can be violated or ignored with impunity if they perceive an inconsistency in the relationship. Even if the inconsistency is created by them through the restructuring of the company or the changing of the personnel involved in the negotiation."

CHAPTER 9
Marketing Transnationally

The stakes in a global environment for technological products are high. According to Geoffrey Moore (*Crossing the Chasm: Marketing and Selling Technology Products to Mainstream Customers,* NY: Harper, 1991):

> High-tech inventiveness and marketing expertise are two cornerstones of the U.S. Strategy for global competitiveness. We ceded the manufacturing advantage to other countries long ago. If we cannot at least learn to predictably and successfully bring high-tech products to market, our counterattack will falter, placing our entire standard of living in jeopardy. (Moore, 4-5)

Marketing for most engineers is a daunting practice — part mumbo jumbo, part statistics, part wishful thinking, part hype. It is the art of persuasion through the emotions. Persuasion for engineers is generally through logic, reason, and numbers. No one expects you to be a marketer in your own country, let alone another one. But understanding the power of the strategy to bring high-technology products to market can make a perplexing environment into a familiar experience.

The goal of any serious marketing effort is to convince consumers at any level that the products and services you offer are better, cheaper, faster. In other words, quality, price, and performance are the fundamental concerns people have in buying technology products and services. They may, of course, trade off lower quality for lower price, or higher price for higher performance. Marketing helps determine when trade-offs occur.

Let's look briefly at the kinds of people who buy technology products. According to Moore (10-14), the five groups are: 1) Innovators; 2) Early Adapters; 3) Early Majority; 4) Late Majority; 5) Laggards. Each group represents a unique *psychographic* profile. Each is different in psychology and demographics, making the responses of each group different from the group next to it.

1) Innovators:
- pursue new technology products
- seek them out before marketing begins
- place technology at the center of their lives
- take pleasure in exploring new technology for its own sake
- make up a small, but influential minority; their endorsement counts

2) Early Adapters:
- buy new products early in the life cycle, but not technologists
- easily imagine, understand, and appreciate the benefits of the new technology
- relate potential benefits to their own goals
- make their buying decisions on their own intuition rather than another's recommendation

3) Early Majority:
- share some of the early adapter's appreciation of technology, but driven by practicality
- wait to see how others do before they buy in since they know how fads work
- require well-established references before they invest
- make up about 1/3 of the adoption cycle and are critical to the success of any product

4) Late Majority:
- shares all the concerns of the early majority, and more
- possesses no comfort with the ability to handle a technological product
- waits until a standard is established

- requires lots of support from a large, well-established company

5) Laggards:
  - want nothing to do with technology for many reasons, personal and economic
  - buy technological products buried deep in another product so it is invisible to them, like the microchips in an automatic coffee maker
  - are generally not a target for high tech approaches

Moore offers a technology adoption life cycle (Figure 9.1)

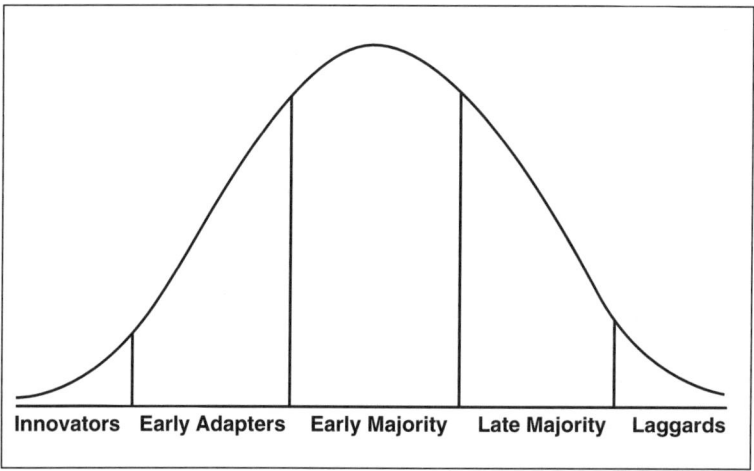

FIGURE 9.1 Moore's Technology Adoption Life Cycle
Source: Moore, Crossing the Chasm

According to Moore, the marketing model moves along the curve from left to right, first capturing the innovators and growing that market, then moving to the early adapters and growing that market, to early majority, late majority, and then laggards. It seems easy. Be first in the market with a product, catch the curve, and enjoy a monopoly. Sound too easy? You are right. The breaks between the groups are like thresholds between changes of state — liquid to solid to gas. Getting over the threshold or, as Moore describes it, across the chasm.

Each culture will share some of this progress from group to group. And ironically, some groups of professionals will share more with their counterparts in other countries than with people of their own country and culture. So in selling a new software package to a configuration data management professional in Moscow, keep in mind that person shares more, on this subject at least, with his or her counterpart in London than the office assistant who speaks the same language.

The approach you take to marketing an item may often be driven by your own company's culture. In marketing worldwide, the debate continues whether a particular campaign can be used in several countries as is. The other side of the debate argues that no product marketing or advertising travels across cultures without some serious examination of its appropriateness for the context. It is the caveat: "Think global, act local."

However, some products, particularly category leaders, can transcend local perceptions and be recognized as leaders across the globe. Recognition leads people to buy the brand they see as the leader. As Europe attempts to create a single market and blur the borders across the continent, understanding how a product travels is important as well. The London office of DMB&B (D'Arcy Masius Benton & Bowles) says that a brand leader has any one or a combination of the following:

- owns/defines product category
- largest
- quality
- best
- longevity
- range
- highest profile

A leader in a category establishes a relationship with a customer that is one of three types:

- rational trust — enables the customer to do something and respond to the brand by thinking "I agree it does/ I need to do this"

- personal identification — enables the customer to be something and respond to the brand by thinking "I am this/ I love being this"
- ideological dream — enables the customer to belong and respond to the brand by thinking "I want to be part of this (group/movement)"

These three categories operate as Power or doing, Identity or bonding, or Icons or belonging.

Power leaders are brands that offer superior evidence of benefits in use. Continual innovation and R&D are necessary to achieve and maintain this power status. As such, a power leader in a category travels well, if the customer criterion is the same from country to country. If so, the strategy and the execution can be transported from country to country.

Identity leaders have formed a personal, emotional bond with the customer. The customer identifies with the "personality" of the brand. The customer gets emotional security in return. Identity leaders travel across borders if you are sensitive to differences in personal goals. Any advertising or marketing themes need to be adapted to the emotional bonds in different cultures. The strategy may travel across borders, but the execution needs to be adapted to the culture.

Icon leaders are rare, but successful and powerful. The advertising for such products is not the product or the user, but "the world" that is associated with the brand. Such scope requires "big" advertising to create the symbolism. Coca-Cola, Levi's, Kodak, British Airways are examples of icon leaders. Since icon leaders deal with universal dreams, they travel extremely well.

CHAPTER 10
Managing People Globally

Successful managers practice "people" skills, as well as technical excellence. Keld Alstrup, Vice President of Human Resources for Volvo Cars of North America offers a strong, bottom-line rationale for companies to pay close attention to the human side of doing business globally — organizations spend more on dissatisfied employees than on satisfied ones, and satisfied employees are more productive. Volvo has an expatriot "pre-visit" to prepare the employee and family before they take the job. The company requires that the entire family commit to the move, to have ownership of the decision to live and work in another country. They advise engineers to treat the overseas assignment as a project that must be managed. One piece of advice for engineers in managing others in a foreign country is to remember that cultures and customs differ. Keep in mind that what others do "is not wrong, it's just different." Follow the locals and you will not go wrong. As an expatriot or foreigner, you have to be in line with the host country.

People skills are essential for successful management anywhere in the world. Thom O'Connor of Price Waterhouse reminds us of the conditions necessary for effective interpersonal relationships:

- encounter each other personally, meeting on a person-to-person basis
- empathize accurately with each other's private world and communicate that understanding significantly to each other
- regard each other warmly and positively despite the particular behavior of either party at a given moment

- regard each other positively and unconditionally, without evaluation or reservation
- perceive that a mutually maintained open and supportive climate reduces the tendency to distort meaning
- exhibit trustful behavior while at the same time reinforcing feelings of security about each other

These conditions are useful for anyone who wants to communicate with any degree of success on a personal or professional level, at home or abroad.

Another essential element in managing successfully is language. Chapter 1 began our discussion of working in a global environment with the importance of language. Here we can add the importance of English as the international language of engineering and science. An international audience certainly knows English, but it is important to keep in mind that the majority of your audience will be using English as their second language. Speaking and writing in English for an international audience requires that you pay special attention to the words you choose, the length of the sentences you create, and the figures of speech you use to illustrate a point. Figure 10.1 offers five types of English, the goals of each, and some characteristics that should prove helpful.

Also as we have mentioned, the class and economic structure of a country influences the way you manage in that country. For example, in most U.S. based corporations that are involved in the research, design, development, and manufacture of high-tech electronics, the management style has changed over the past 10 - 25 years. Management after World War II was hierarchical and authoritarian. Managers commanded and subordinates obeyed and responded.

That old-style approach has been replaced with work-teams empowered to do work on their own. Such decentralized management practices work in the U.S. because the workforce thrives on an approach that recognizes the emphasis on individualism and freedom. As we have discussed above, some cultures, see Figures 2.3, 3.2, and 5.2 do not always fit

KIND OF ENGLISH	GOAL OR PURPOSE	CHARACTERISTICS
Basic English	Aimed at easy social communication by reduction of vocabulary	
Controlled English	Designed for non-native and native speakers with a low level of education. Aimed at technical communication (instruction and maintenance manuals)	No synonyms allowed. One meaning per word (whether technical or not). Simple, standardized constructions. Restricted use of verb tenses, irregular verbs, auxiliaries. Use of active voice only. No idioms, contractions, abbreviations, acronyms, slang, or jargon
Simplified English	Designed to avoid the need for translation. Similar to controlled English. Developed for the aerospace industry	No synonyms allowed. One meaning per word (whether technical or not). Simple, standardized constructions. Restricted use of verb tenses, irregular verbs, auxiliaries. Use of active voice only. No idioms, contractions, abbreviations, acronyms, slang, or jargon
Plain English	Similar to controlled English and simplified English. Developed by Unisys Corp. for the computer industry	No synonyms allowed. One meaning per word (whether technical or not). Simple, standardized constructions. Restricted use of verb tenses, irregular verbs, auxiliaries. Use of active voice only. No idioms, contractions, abbreviations, acronyms, slang, or jargon
International English	Designed for an international audience of technical experts. Used for scientific, abstract, or mathematical materials. Designed to facilitate translation	One-meaning-per-word rule applied to technical terms only. Clarity achieved through careful grammatical constructions and word choices. Idioms, jargon, abbreviations, and acronyms avoided

FIGURE 10.1 Five Types of English and Their Uses for Writing and Speaking to International Audiences

Source: Los Alamos National Laboratory, International Communication Committee

such new approaches. The culture of the country, the class of the managers and employees and the norms of the profession in a particular country influence heavily the success of a particular management style.

Often managers have more in common with other managers and professionals from different countries than they do with people from their own nation who are not of the same class or profession. Figure 10.2 shows this difference. Keeping such class, country, and profession differences in mind can help you manage a multinational and multicultural workforce where ever you are, at home or overseas.

The Manager's role in communication in a transnational setting, as companies adapt to the enormous global forces driving business, is becoming more and more the responsibility for the person on site. The path to effective communication within an organization tends to be rather simple even in an environment of global complexity. The simplicity of successful internal communications is based on a company culture that emphasizes and practices two-way communications: managers,

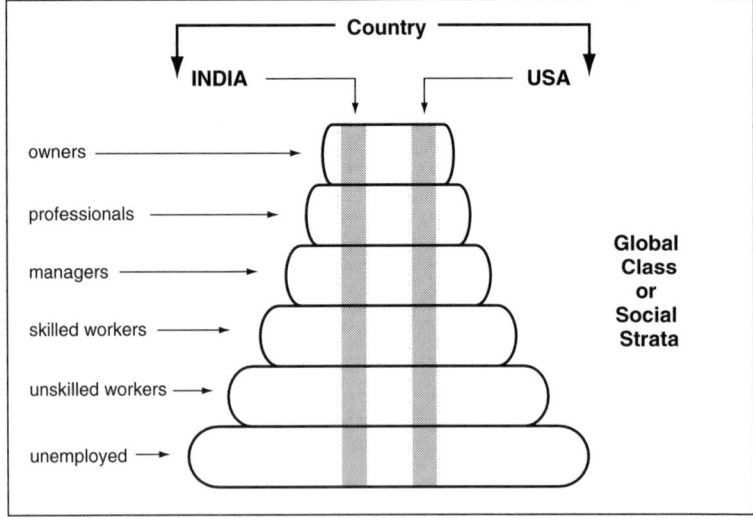

FIGURE 10.2 People Often Share More with Members of Their Own Professions Than with Fellow Countrymen

supervisors and directors not only talk to and inform the work-force about day to day matters, events and developments, and company news, they must also listen and gather feedback from the workforce. They must act on the feedback to encourage employee involvement. You can encourage employee involve-ment through effective two-way communication, encouraging employees to put forward ideas and suggestions, giving employees responsibility for solving problems and implement-ing improvements, setting up work teams.

A company culture of open communications and employee involvement demands that managers have the confidence to share information, to encourage feedback, and to give employees responsibility so that they can become more involved in their work and feel they are making a valuable contribution.

An effective flow of information up down, and across levels in an organization is essential in any successful operation anywhere in the world. The most credible, effective, and valuable method of communication is face to face. Frequent and regular meetings and encounters are the key. A variety of other methods and media are also available. Figure 10.3 is based on a chart of employee communications used at the Peugeot Talbot Motor Company PLC, the joint French and English manufacturing venture in Coventry UK.

It may be appropriate to mention here that global manage-ment skills may be required even if you have never traveled more than a few miles from home. Electronic communica-tions, digital media, fax machines, company computer net-works, and the Internet are the tools used daily in the "glob-al village" of business. The world community Marshall McLuhan predicted more than a quarter century ago in *Understanding Media* is the workplace of engineers, lawyers, scientists, bankers, and educators. The global workplace has a voracious appetite for information and information technol-ogy. The etiquette of on-line communication, contrary to much of what has been written, should be no different than writing an internal memo or report, or a letter to someone

METHOD	PURPOSE	RECIPIENTS	SPEED	FREQUENCY
Managers & Supervisors				
-Company Charter	Mission & Vision Statement	All Employees	2 or 3 days/ shifts	Monthly
-Company Operations News, Briefing Meetings	Operations Committee Meeting - functional directors; Briefing Document with details on productivity, sales, finances	All Employees		
-Urgent Information, Communications -Immediate Document	To get very urgent information to all employees before the media. Known as "Red Documents"	Functional Heads, Line Mgt & Supervisors, then verbally to all other employees	Information reaches all employees the same day	As required
-General Information, Communications General Document	Less urgent news, but can't wait for next company newspaper issue. Known as "Green Documents"	All employees verbally from supervision or posed on notice board	2 or 3 days	As required
Manufacturing operations				
-Local Operations News, Manufacturing Supervisor's Brief	Snap-shot of events at plant from t he previous week; Production achieved, quality, targets; detailed work area information	Manufacturing Supervisors	1 day	Weekly
-In House Radio, Radio Ryton	Factory areas - music sports results, instant communication on non-contentious items	Shop floor employees	Immediate	Daily Broadcasts
Overview for management				
-Company Overview	Managing Director's presentation - evening meetings with all managers, or in small groups; Trade Union reps meeting	Management; Senior Trade Union Reps.	2-3 weeks	Once or twice a year

METHOD	PURPOSE	RECIPIENTS	SPEED	FREQUENCY
Company Wide -In House Newspaper, "Times" Color Tabloid	Newspaper for employees blending employee, industry and company news	All employees; circulated to Company dealers, Press, Libraries, Retirees, Other Companies	1 or 2 days from publication	10 issues per year
-Training & Communication Videos	Company produced video on changes in operations, facility, quality, pension scheme, etc	All employees, Financiers/Bankers, media, suppliers, community		
-Annual Report	Report on company performance and accounts	Sited throughout all plants and offices	Within 2 weeks of publication	Annual
-Notice Boards	Display of detailed information on negotiations, job vacancies, fire/safety regulations, sporting and social events	All employees	Distribution to all Notice Boards in 1 or 2 days	As required
General -Public Relations/ Marketing Publications	To promote launch of new car and emphasize commitment to Coventry	Management Trade Unions Notice Boards	Within one week of printing	As required
-Agreed Minutes	Jointly agreed Minutes of negotiations on Pay and Conditions containing Trade Union submissions and Company offer	Appropriate group of employees	Immediate	During Negotiations As required
-Direct Contact	Company communicates directly with employees by mail, primarily when faced with a potential or actual dispute but also on other occasions		Distributed by mail to employees' homes or handed out	

FIGURE 10.3 Frequent and Regular Employee Communications Similar to Peugeot Talbot's Are Common Among Transnational Corporations

Source: Peugeot Talbot Motor Company PLC

outside the company. Business communication, on paper or
on-line, is a reflection of your company's and your own pro-
fessionalism. Face to face or terminal to terminal clear, con-
cise, and cogent messages are the mark of successful and
effective people and organizations.

Issues of Leadership, Power, and Authority. Americans
overseas often assume that their office behavior at home
translates well abroad. In many technology-driven industries,
particularly electronics, a "shirt sleeve" environment prevails
around the office. Such dress-down behavior violates the con-
ventions of rank and power in many parts of the world, such
as Europe, South America, and Asia. Appearance, clothing,
the size of the office, titles, as we have discussed above pre-
viously, have different meanings and different importance
throughout the world. We may live in a global village, but
local customs and habits in business practice should not be
ignored. In most countries your mere presence signals your
authority and power. Americans raised in a democratic society make a large effort to appear like one of the people. Also,
Americans, as we have noted, have difficulty with the clear
class distinctions they see in Latin America, Asia, the Middle
East, and Africa. Such class and power distinctions should be
evident to you in interpersonal contact, conversations, and
body language.

Power does not necessarily mean a lack of relationships. In
many hierarchical countries, a "personalized style" translates
into appearances at birthday parties, soccer games, walking
through work areas, talking and listening to workers, calling
them by name, asking them how they are doing without men-
tioning work.

In short, Americans often see the workplace as the place of
work; in other countries work is often intertwined into other
aspects of the society at large. Leaders in business are often
leaders in society, the arts, and philanthropy as well.

Decision Making and Delegation. In most countries the way
to decide on issues and the way to tell others to do things is

not the same. The differences lie on a continuum: see the power/distance chart above Figure 5.1. At one end is an authoritarian, centralized approach, and at the other a Japanese-style participatory concensus-making approach. American managers are roughly in the middle, along with most Scandinavians and Australians. As corporations adopt more and more global practices you will encounter similar practices involving work teams, Total Quality Management principles, and decentralized decision making.

Work Ethic, Supervising, Hiring, Firing, and Motivation. A range of incentive and reward options to motivate a world-wide workforce is needed to meet the personal goals and aspirations of individuals in different countries over time, and in different stages in their careers. Observe individuals, talk with them, find out what makes them work hard, then select the incentive or reward that best fits the person: money, vacation time, personal respect, public recognition, family security, job challenge, advancement, title, power, social acceptance, gifts, access to sports or health facilities, services, reserved parking, big office, staff, and so on.

CHAPTER 11
Dealing with Culture Shock

- Homesickness
- Boredom
- Withdrawal (spending excessive time reading; only seeing other Americans; avoiding contact with host nationals)
- Need for excessive amounts of sleep
- Compulsive eating
- Compulsive drinking
- Irritability
- Exaggerated cleanliness
- Marital stress
- Family tension and conflict
- Chauvinistic excesses
- Stereotyping of host nationals
- Hostility toward host nationals
- Loss of ability to work effectively
- Unexplainable fits of weeping
- Physical ailments (psychosomatic illnesses)

Source: Kohls, Survival Kit for Overseas Living

That is a list of some of the symptoms that you may observe in relatively severe cases of culture shock. An overseas assignment can be stressful, and the stress of being in another culture is known as culture shock.

Dealing with culture shock, at its most basic is dealing with *change.* In dealing with change we focus on people, expectations, and culture. As we mentioned, some people are suited for work outside their own country. And keep in mind that *all change is personal.* If we build on that observation, then we can ask these questions: What type of person is best suited for change? What role do expectations play in the change process? Do some corporate cultures adapt to change better than others?

What type of person is best suited for change? You may have noticed around your organization an individual, and that person may be you, who sees the widespread changes in work processes and outcomes as a stimulating challenge. This person comes to work with a smiling face and a spring in the step, often arriving early and leaving late. No matter how much chaos the organization is in, this person appears to respond well to the situation.

Others in the organization respond less well to change and exhibit dysfunctional behavior. There are degrees of dysfunctional behavior related to change. For instance, examples of a low degree of dysfunction are: poor communication, reduced trust, blaming, defensiveness, increased conflict with fellow workers, decreased team effectiveness, inappropriate outbursts at the office. Moderate dysfunction: lying or deception, chronic lateness or absenteeism, symptoms such as headaches and stomach pains, apathy, interpersonal withdrawal. A high degree of dysfunction: covert undermining of leadership, overt blocking, actively promoting a negative attitude in others, sabotage, substance abuse, physical or psychological breakdown, family abuse, violence, murder, suicide.

The person who responds well to change exhibits buoyancy, elasticity, resilience — the ability to recover quickly from change. Note such people possess a strong, positive sense of self which provides them the security and confidence to meet new challenges, even if they do not have all the answers. These people, like successful athletes, are focused on a clear vision of what they wish to accomplish, and they are tenacious in making the vision a reality. In addition, these people tend to be accommodating and flexible in the face of uncertainty, and organized in the way they develop an approach for managing ambiguity. These people are proactive. They engage the circumstances, rather than defend against change.

The type of person I have described here is not that unusual. Such a person practices fairness, integrity, honesty, and human dignity — the principles that provide us all the security to adapt to change. Such people do well in any environment, particularly in a foreign country.

What role do expectations play in the change process? If, as I have suggested, all change is personal, then how can understanding and managing expectations help an individual or an organization through the change cycle? Everyone has made personal changes: leaving home for college, getting married or divorced, relocating to another town. Each personal change brings with it the feeling that *things* will get better. Like most of the characters in Dickens' *Great Expectations,* fame and fortune are often illusive and illusory because they neglected to consider that change is an equal opportunity for failure. A contemporary rock song puts it this way, "If you don't expect too much from me you may not be let down."

However, it is not to lower expectations, but to *manage* them. In managing expectations, consider that in responding to positive change most people go through phases: 1) uninformed optimism or certainty at the start, like the joy at a wedding; 2) informed pessimism or doubt — here people may quit publicly, or more destructively they will quit privately and continue to work, allowing the negative feelings to generate dysfunctional behavior; 3) hope emerges with a sense of reality; 4) informed optimism results in confidence; 5) satisfaction closes the cycle of change upon completion.

The good news is the cycle is predictable and can be used to manage expectations by helping people prepare for the rough periods. The bad news is most people feel they are the exception and they will not follow the cycle from beginning to completion.

Do some corporate cultures adapt to change better than others? The concept of corporate culture is complex, but for this discussion, we can consider that it is made up of the physical things and patterns of behavior that reflect the values and beliefs and basic underlying assumptions of the corporation or organization. A culture that values the *status quo* may resist change, but may paradoxically be best suited to meet the challenge of change. A PROCESS culture such as a public utility or telecommunications company, may have the scope and resources to make a successful cultural change. It has the capacity to survive as the people and processes go through the cycle of change. A MACHO culture such as an investment bank or movie studio, may be entirely wiped out by changes in laws or in the economic environment. AT&T & IBM are still alive, while E.F. Hutton and Drexel Burnham are not.

The survival of an organization, like the survival of an individual, also depends on its buoyancy, elasticity, resilience — its ability to recover quickly from change. Corporations which have such people have high organizational abilities, and they do well in other countries.

PART II

UNDERSTANDING AND WORKING IN SPECIFIC REGIONS AND CULTURES

- Europe and the European Community
- Eastern Europe and the Former Soviet Block
- South America
- NAFTA
- The Pacific Rim

CHAPTER 12
Europe and the European Community

Europe has historically been the site for most overseas work. The nations of Europe have been and are global partners, adversaries and leaders in world trade. No matter what part of the world you are from, Europe figures into the mix.

Doing business in Europe would seem on the surface a no-brainer. History, culture, and often language all point toward an equitable arrangement. But, pay special attention to the differences, for the misconception of similarities like the subtle differences among members of the same family, can often grow into disputes and blocks to economic partnerships. Take for example the case of a Detroit-based diesel manufacturer's European CEO. The person had been successful in Detroit and as a teacher in Cairo, but had an abrasive style. He delegated responsibility to set up a program in Zurich to create a European team.

> The Europeans call him an *Arbeitstier*—a "work animal"—because he always works late. He never joins the staff for a leisurely lunch, preferring to eat a sandwich at his desk. He still can't speak even rudimentary Swiss German... On top of that [his] wife is unhappy in Zurich because she misses her job, and their daughter is upset because she's having trouble applying for U.S. colleges from the school she's attending. (Adler, Gordon, "The Case of the Floundering Expatriate," *Harvard Business Review* July/August 1995:166.)

This highlights, according to Fons Trompenaars author of *Riding the Waves of Culture* (1993), a common misunderstanding of the challenges involved in managing in Europe.

Creating an effective team there means more than helping people adjust to another business model. It requires commitment and the selection of people who are not only excellent technically, but who also understand the organizational and behavioral forces at work. Europeans manage by objectives. They apply total quality principle and survey their employees and customers. But often the person sent from the United States to do these things is a subject matter expert, with little or only a surface knowledge of the context of the foreign environment. Working together overseas requires a deep understanding of the differences in local cultures. "It's amazing what can be achieved when one makes a business issue out of intercultural experiences. Increasingly, international managers realize that they can gain competitive advantage by understanding cultural differences. Technologies can be copied quickly. Intercultural competence cannot be copied; it must be learned." (*Harvard Business Review,* July-August 1995:38.)

In understanding cultural differences and working in Europe and the European Union consider:
- the impact WWII on Europe
- the nature of work and their bureaucracy
- the economic nature of the European Union and its policies
- their attitudes toward business activities
- the rules and regulations necessary to do business there
- the inherent political tensions

THE IMPACT OF WWII ON EUROPE

Wars, particularly recent ones, play a profound role in shaping the way Europeans think, and by extension, the way they think about business. The absolute devastation and destruction brought on by World War II left almost the entire continent in ruin, financially exhausted and in despair. Leaders within Europe, and throughout the world, felt that to continue the cycle of war and destruction in an atomic age put the entire world at risk.

On the other hand, a Europe at peace would point to prosperity. One approach to forging peace in an era of increasing military tension brought on by the Cold War with the Soviet Bloc was through economic cooperation. In April 1951 the European Coal and Steel Community was established between Germany and France. Placing these adversaries in cooperation over the raw material of war was to be a model of cooperation for other areas. Figure 12.1 offers an overview of important treaties and events that led to the evolution of the European Union.

Membership in this trading and economic partnership began

DATE	EVENT
1951 1957	Signing of the Treaty of Paris establishing a European Coal and Steel Community (ECSC) Signing of the "Treaties of Rome" establishing the European Economic Community (EEC) and the European Atomic Energy Community (EAEC)
1968	The Community becomes a customs union, import and export duties abolished among members
1973 1979	Denmark, Ireland, and the UK join the EC The European Monetary System (EMS) comes into operation First direct elections to the European Parliament
1981 1984 1986 1989	Greece joins the EC Second direct elections to the European Parliament Spain and Portugal join the EC The Single European Act amends the Treaties of Paris and Rome Third direct elections to the European Parliament
1990 1993 1995	Germany reunited after the fall of the Communist regime in the East Treaty of Maastrict on European Union sets goals for a frontier-free Europe, economic and monetary union, and social and political union Austria, Finland, and Sweden join the European Union

FIGURE 12.1 Events Since WWII Have Resulted in the European Union as a Major Economic Entity
Source: Various European Community Publications

in 1959 with Belgium, France, Germany, Italy, Luxembourg, and The Netherlands
- 1973, Denmark, Ireland, and The United Kingdom
- 1981, Greece
- 1986, Spain and Portugal
- 1995, Austria, Finland, and Sweden.

The list could expand, since Hungary, Poland, Cyprus, Malta, and Turkey have applied to join. And the end of the Cold War has prompted the EU to draw up plans to include in its single

FIGURE 12.2 The European Union has Evolved into More than an Economic Union to Become a Global Social and Political Force

Source: The European Union, New York

market Bulgaria, Czech Republic, Hungary, Poland, Rumania, and Slovakia. Other possibilities include association agreements with Estonia, Latvia, and Lithuania. Figure 12.2 is a recent map of the European Union.

THE NATURE OF WORK
AND THEIR BUREAUCRACY

Now you might be wondering why so much discussion of history and politics? Because, key to understanding local actions in Europe is understanding not only the history and political and social forces at work there today, but also over the course of the previous centuries. That's right—centuries. In some countries like England, the tradition of democracy goes back to the Magna Carta, whereas Germany's current democratic structure is based on America's and is the result of the American influence after WWII.

Also, in many technically based companies, and in most of the Western world for that matter, fundamental shifts are taking place in the way people, organizations, and governments operate. Downsizing, reinventing, reengineering, and restructuring are replacing hierarchical structures and process cultures. Organizations are focused on the work and the outcomes of work, rather than the perpetuation of the structures of work.

Bureaucracy, a French word, has the world over become synonymous with delay, endless paperwork, diffused authority and responsibility, frustration, and added expense in time and money. No matter how much the world is streamlining processes and government, bureaucracy remains a fact of life. It may be a little more responsive and friendly in some countries of Europe, and more efficient and helpful in others, but it is still the way things get done in Europe.

It is a good idea to learn the regulatory bodies and organizations in the country you are in and treat their rules as you would the rules and laws in your own country. Obey them, and if you think they are unfair, then join others in your industry to influence a change. Bureaucracy responds well to groups of organizations and countries. Remember that a nation that has a rich and long history sees change from a much different perspective. Keep in mind that the current negotiations on the Uruguay Round of the General Agreement on Tariffs and Trade (GATT) began in 1987 and the implementation of the agreements started in 1995. The goal is to discipline trade-distorting practices such as tariffs and subsidies.

THE ECONOMIC NATURE OF THE EUROPEAN UNION AND ITS POLICIES

Europe has traditionally been the largest market for the export and import of U.S. goods and services. When compared with Japan in 1992, according to the U.S. Department of Commerce, Europe imported almost $200 billion and exported slightly less to the U.S. for a balance of +$3.5 billion. By contrast, U.S. exports to Japan in the same period were approximately $75 billion, with much greater imports, yielding a balance of -$38.3 billion. Put simply, trade with Europe is vast and for the most part equitable.

Though the relationship with Western Europe is profitable and friendly, care must still be taken in doing business there. Current trends in Europe toward a European Union and a Single Market within that Union have shifted power away from any single dominant country in Europe to the structures of the EU.

The first step in creating a single market was to break down frontiers so "goods, services, people, and capital move unhindered across" the borders of member countries. To bring about the Maastrict Treaty, aimed at removing the physical, technical, and fiscal barriers to trade. Achieving this goal has come a long way, and is most evident in the border crossings which now funnel EU citizens one way, and all the rest another.

Internally the efforts for deregulation and reform continue within the EU. To have regulations on the quality of goods and services uniform throughout the EU has fallen largely on the committees and working groups within the political arm of the EU.

Figure 12.3 shows the relationship of member governments to the EU. Knowing the layers of political power in Europe is fundamental to your understanding of the power of any regulating body there. In Europe, agencies or committees or working groups set the policy and stipulate the standards for such things as electrical appliances, videotape, or the purity of beer.

FIGURE 12.3 Working in Europe Demands an Understanding of the EU as Another Political and Bureaucratic Force

Source: Working Together—The Institutions of the European Community. (EC: Brussels, 1994)

RULES AND REGULATIONS NECESSARY TO DO BUSINESS THERE

Regulations such as ISO 9000 have an enormous impact on corporations doing business with the members of the EU. If your organization does not meet ISO standards, you cannot do business in any EU member country. Think of it. Miss the standard, and your product or service, no matter how good, cannot find its way into what is now one of the three largest markets in the world, and if the expansion continues as we mentioned above, the world's largest market.

Those standards are developed in working groups that are part of the permanent bureaucracy of the EU. Organizations of international groups such as the IEEE often advise and contribute to the setting of technical standards for the EU. Increasingly, these ISO standards are becoming the world standard, which is another example of how the economy has become more and more global.

US and Japanese companies stand to gain from the efforts to create a single market in Europe since both are better at and more flexible in restructuring their operations. For instance, General Motors can now consolidate its automobile operations by designating one site for electrical parts to serve all of its EU operations and replacement parts, rather than having sites in each member state as before.

Along with the advantages for those companies who think globally and make the effort to work with their European partners and customers, some systemic problems remain for US business. For instance, gaining access to power. In short, who do you lobby? Other systematic problems include: creating effective organizations that work in the EU; gathering accurate information; becoming an effective player in a changing Europe; striking a balance between marketing at national and EU levels.

Specific industrial sectors must meet the challenge of changing standards, testing, and certification. Environmental and consumer protection issues need to be addressed, as well as the changes in public procurement. Industrial policy, competition, and subsidies, as well as social issues and work rules add to the considerations in doing business in Europe.

EU ATTITUDES TOWARD BUSINESS ACTIVITIES

Business attitudes in Europe are different than ours, but, as was shown in Figure 10.2, people with similar educational, class, and professional backgrounds tend to share a great deal internationally. To support the notion that you share more with an engineer from Coventry, England or Frankfort, Germany than you do with an auto mechanic in Minneapolis, let's look at *Tomorrow's Company: The Role of Business in a Changing World* (London: RSA, 1994). This report by the Royal Society for the Encouragement of Arts, Manufactures, & Commerce (RSA)—the British equivalent of The Conference Board—challenges business to meet worldwide competition through an inclusive approach.

> In an inclusive approach success is not defined in terms of a single bottom line, nor is purpose confined to the needs of a single stake holder. Each company makes its own unique choice of purpose and values, and has its own model of critical business processes from which it derives its range of success measures. But tomorrow's company will understand and measure the value which it derives from all its key relationships, and thereby be able to make informed decisions when it has to balance and tradeoff the conflicting claims of customers, suppliers, employees, investors, and the communities in which it operates. (*Tomorrow's Company: The Role of Business in a Changing World*: 1)

This maintains that the forces of global competition are making change necessary. Other complicating factors are rising population and consumption putting pressure on natural resources; rapid changes in technology changing employment patterns; changes in people's aspirations, the rise in pressure groups, and reduced public confidence in governments and other institutions. To compete internationally, a company needs a supportive operating environment. A shared vision and common agenda among business, government, and the community is key to meeting the challenge of competition.

Winners in this international arena maintain their "license to operate" by achieving a high level of support from everyone they contact directly or indirectly. Such companies must learn and change rapidly, inspire new levels of skill and creativity in their people, and develop a shared destiny with customers and stakeholders. Traditional measures of success, such as financial performance and returns to shareholders, must not continue as the only purpose of business. All a company's relationships must be included in its definitions and measures of success. The report emphasizes that tomorrow's company has clear values and purpose, defines relationships consistently, is part of a wider system, recognizes its dependence on relationships, recognizes the need for trade-offs among stakeholders, understands the need to measure and communicate its performance in all its relationships.

If the RSA (Royal Society for the Encouragement of Arts, Manufactures, & Commerce) report sounds familiar, it is because the issues and the dialogue are about global business. These concerns transcend borders and geographic regions. They recognize, as Europeans have for centuries, that the world is a small place. Relationships and partnerships are needed for success.

THE POLITICAL TENSIONS
THAT ARE INHERENT THERE

Finally, many non-European capitalists continue to be surprised by the relative harmony between business and government. Compared with the almost adversarial relationship found in the U.S. Europe has had a long mercantilist tradition in which the interests of private companies and the state are in harmony. Add to this the prominence and maturity of government established long before the rise of big business and industrial organizations. Bureaucrats saw no threat since they were charged with the public welfare. By contrast, big business appeared in the U.S. before big government, and relations are often strained and even hostile.

The Maastrict Treaty and the movement toward a single market are manifestations of the European effort to remove some of the inefficiency, cost, poor service, and stodgy habits that have emerged from comfortable relations between business and government. But, to compare capitalism in Europe and the U.S. only in terms of economic efficiency and performance, would lead to misunderstanding.

> Given a choice between better profits and higher dividends on one hand, and the social stability represented by high employment rolls on the other, Europeans have usually chosen stability. Up to now, [they] have been willing to accept the overhead burden this choice imposes by paying higher prices and accepting lower returns... Economic outcomes, for both companies and individuals, are more tightly "bunched" than they are in the United States and Japan. And most Europeans would sacrifice the possibility of an unrestricted business environment that rewards a few with extreme wealth for the reality of many more people with comfortable incomes. (Henzler, Herbert. "The New Era of Eurocapitalism," *Harvard Business Review* July-August 1992:62.)

The history of a Europe with wars, revolutions, and vicious labor disputes helps a non-European begin to understand why they emphasize stability. Their history of the long-term costs of unrest and violence make any short-term price seem a bargain.

CHAPTER 13
Eastern Europe and the
Former Soviet Bloc

During the Cold War writing a section on doing business in Eastern Europe and the Soviet Union might have been considered un-American, or at least an oxymoron since communism and capitalism literally fought for world dominance. A managed economy and a market economy had little in common.

Efforts to establish a market economy in the aftermath of the collapse of the Soviet Union have created enormous opportunity, as well as tremendous upheaval. To the credit of the Russians, the 1991 revolution was not violent. However, it did have considerable social and economic impact on the lives of Russians and on the people in the 15 successor states of the Former Soviet Union. Figure 13.1 is a map showing the location of the 15 Soviet states. This split has also had a strong impact on the former Communist countries of Eastern Europe, particularly Poland, Bulgaria, the current Czech Republic and Slovakia, Hungary, and Rumania.

To understand cultural differences and working in Russia, Eastern Europe, and the Former Soviet Union, one must consider the impact of WWII (1945) on Russia and Eastern Europe, the nature of work and their bureaucracy, their attitudes toward business activities, rules and regulations necessary to do business there, the realities of establishing a market economy, and the political tensions that are inherent there.

IMPACT OF WWII (1945) ON RUSSIA, THE FORMER SOVIET UNION, AND EASTERN EUROPE

When working in Russia always remember that World War II continues to have a significant role in how Russians, not just Russians of that generation, but how all Russians think. For them the victory over Nazi Germany saved the world, and that price was paid dearly with the blood of literally millions of Russian soldiers and civilians. Theirs is a mindset similar to the British about the price they paid to defeat Hitler and save the world from Nazi domination. As we mentioned in the previous chapter, history is a powerful force in Europe.

Russian history is also filled with political, military, and cultural conquests of their country by Western Europe. An uneasy admiration and disdain for everything western prevails and becomes key to the way Russians must do things, even to the development of a market economy. Placing the U.S. brand of capitalism in the heart of Moscow, St. Petersburg or Tallin was attempted in the period just after the election of Boris Yeltsin. The merchant tradition in Russia had been wiped out with the 1917 revolution. The buildings in the section of Moscow owned by traders and property owners were replaced by Soviet offices for the KGB. The legacy of capitalism was literally erased, so it is easy to understand that almost three quarters of a century later, the effort to establish trade and competition would be difficult for people who could not conceive of the private ownership of property much less a business enterprise.

THE NATURE OF WORK AND BUREAUCRACY

Working relationships among technical experts have gone on in spite of political, economic, and cultural turmoil. Technical professionals have exchanged information and engaged in joint initiatives on nuclear power, space vehicles and structures, environmental issues, mining, lumbering, manufacturing, electronics, and telecommunications. However, doing business in Russia is not like doing business anywhere else.

In early 1993, conventional wisdom suggested that the way to do business in Russia was to create a joint venture and use the Russian partner to manage the effort (*Harvard Business Review,* January-February 1993:44-54). That continues to be the most successful formula, but almost all such arrangements with the West have encountered financial disputes and cultural misunderstandings. (*New York Times,* May 6, 1995:35) The most successful ventures in Russia seem to be ones that began almost with the 1917 Revolution, not the 1991 one. Fiat, for example, has been successfully building automobiles in Russia for decades.

THE REALITIES OF ESTABLISHING A MARKET ECONOMY

A Western work ethic has yet to emerge. Old ideas die hard and anyone who has amassed wealth in Russia seems to be a mobster or a member of the criminal sub-culture. Images in early Russian advertising for automobiles, and even the costumes of the acrobats in the Moscow Circus looked like characters off the set of "Bonnie and Clyde" or "The Untouchables." Reports in *Global Finance* (September 1993:48-54), *The New York Times* (July 8,1994:D1; May 6,1995:35) and other media underscore the high level of corruption and organized crime.

The vast difference in living standards is part of the problem as well. The economic downturn in Russia was worse than that of the Great Depression in the U.S. in the 1930s. Inflation as high as 25% a month, a depressed gross national product, an alarming deterioration in the standard of living, bloody ethnic conflict in the Caucasus region of Chechen and Ingush, conflict in Azerbaijan and Armenia, unrest in the Baltic States, and civil war in Tajikistan, have seriously hampered international efforts to stabilize and reform the Russian economy.

Russian workers went from serfdom to communism with the 1917 Revolution. With little or no experience in a market environment, it is easy to understand that Russians stare blankly when Western managers speak of Total Quality Management, re-engineering, management by objective, the learning corporation. They do know, however, how to get things accomplished. Under the old regime they could barter goods and services. A construction project that needed cement, for example, would exchange fuel for the commodity, getting around the system. Doing work with them their way, on their terms, with their people seems to be a successful strategy.

ATTITUDES TOWARD BUSINESS ACTIVITIES, RULES AND REGULATIONS NECESSARY TO DO BUSINESS IN RUSSIA

The changes that have taken place in the last few years in Russia have revealed a need for Russian technical communicators to refocus their expertise and skills in order to enter the global marketplace successfully and competitively.

Rather than dwell on the familiar (Cold War) differences between Americans and Russians, a common ground exists and is growing. We share a mutual interest in the successful entry of Russian technical communicators in the global marketplace. We also share:

- an understanding that technology is central to civilization as we know it, and that the masters of technology have a substantial influence on all activities;
- a belief that technology had a major beneficial impact on the peoples of the world, but with such power comes the potential for large, serious, and potentially devastating influences;
- the notion that technological breakthroughs have pro-

found influences on the nature of work, liberating the traditional intensive physical nature of labor to the emergence of a "knowledge" worker;
- the belief that the global marketplace forces the need for clear and rapid communication across borders, as well as among cultures.

If we can agree on these technical communications issues, then we have a firm foundation for building a gateway to communication in the global market.

Contemporary technical communications in a market economy call for consideration of other disciplines not usually associated with the communication of technical information. Technical communicators throughout the world, in order to be effective and competitive now and into the next century, must consider and master each of the following topics:

- audience analysis
- public relations
- crisis communications
- total quality and change communications
- problem-solving
- marketing communications
- corporate culture
- ethics and technical information
- media relations

Each of these ten topics is a study in itself. Nevertheless, the important thing to remember is that the global marketplace now requires technical communicators to think of these issues. Responsible technical communication in a global market environment implies a partnership among all concerned groups — consumers, employees, the community, business, government. It is up to the individual to develop and perpetuate useful channels of communication among these various groups. (Goodman, Michael B. "The Special Section on Professional Communication in Russia: An American Perspective," *IEEE Transactions on Professional*

Communication 37:2 June 1994:90)

POLITICAL TENSIONS INHERENT IN RUSSIA, THE FORMER SOVIET UNION, AND EASTERN EUROPE

Eastern Europe - Poland, Hungary, Bulgaria, Rumania, the Czech Republic, Slovakia - compared with Russia and the Former Soviet Union, is relatively stable. It has been more compatible to partnerships with the west. Companies such as General Motors, Ford, General Electric, Siemens, Sara Lee, and Cadbury-Schweppes have been successful there. International support, though not trouble free, has had some pay off as a result of 1989 Support for East European Democracy (SEED) Act which authorized the creation of the enterprise funds in Poland and Hungary. These helped private sector development in those countries. Enterprise funds for the Czech Republic and Slovakia were created in 1990 and Bulgaria in 1991. ("Enterprise Funds," GAO Report GAO/NSIAD-94-77, March 1994)

Russia remains a risk. However, the importance of the country, its people, its natural resources, and its technology make it a necessary part of any global strategy. The United States obligated funds to provide assistance and economic cooperation under the policies of the Freedom Support Act (The Freedom for Russia and Emerging Eurasian Democracies and Open Markets Support Act of 1992, Public Law 102-511.) The most visible sign of this cooperation is NASA's joint missions with the Russian space station Mir. A total of 19 U.S. agencies committed $1,045 million in grant programs from 1990 through September 1993.

With its vast resources, its promise of opportunity, its almost "Wild West" absence of authority, Russia offers fertile ground for people who are not afraid of risk. The past emphasis on military spending left consumer needs almost untouched. But success here is limited to those who work with the Russians to do things their way.

CHAPTER 14
South America

Listen to the CEO of a manufacturer of industrial equipment:

> In the lobby of our headquarters—a standard-issue
> office building with four floors of steel and glass— there
> is a reception desk but no receptionist. That's the first
> clue that we are different. We don't have receptionists...
> We don't have secretaries either, or personal assistants.
> We don't believe in cluttering the payroll with ungrati-
> fying, dead-end jobs... We don't have executive dining
> rooms, and parking is strictly first-come, first-served.
> It's all part of running a "natural business." ...we have
> stripped away the unnecessary perks and privileges that
> feed the ego but hurt the balance sheet and distract
> everyone from the crucial corporate tasks of making,
> selling, billing, and collecting.
>
> Our offices don't even have the usual number of walls.
> Instead, a forest of plants separates the desks, comput-
> ers, and drawing boards in our work areas. The mood is
> informal. Some people wear suits and ties: others, jeans
> and sneakers. (Semler, Ricardo. "Who Needs Bosses?"
> *Across the Board* February 1994:24.)

Does this sound like a flat, lean-and-mean, cutting-edge, high-
tech firm from Silicon Valley? If you thought so, get set to
discard some of your preconceived notions of how business is
done in Brazil. That's right, the company is Semco S.A. head-
quartered in Sao Paulo, Brazil. If you are headed for South
America, get set to put those images of Banana Republics,
mañana, and *siestas* in with other relics of the past.

Businesses there, and in Brazil in particular, can be among the most innovative, global, and forward-thinking operations in the world. And as mentioned before, the professionals and managers there share a great deal with you. Many of them studied management at U.S. universities, or use management and professional techniques you are familiar with. Similarities among the nations of South America are great: language, religion, history, geography, weather, customs. Spanish is the language of all the countries except for Brazil, which is Portuguese. And the predominant religion is Roman Catholic.

However, in doing business in South America, keep at the top of your list that each nation of the continent has its own distinctiveness. Understanding differences among the nations there is extremely important. A good overview of the country from its embassy, or from your own country's embassy, can be a fine start. Some familiarity with the country's history and heroes, its local folklore, and its most revered writer or artist can give you a window into the way your hosts and business associates view the world and you as a foreigner. It also signals that you are interested in them, as we have mentioned over and over in Part I, and that you have taken the trouble to know more about the country you are in and its people.

Some generalizations can be made about common business practices. In much of Central and South America the custom is to shake both hands when you arrive and when you leave. Hugging among acquaintances is common. Conversational distance is close, and eye contact tends to be constant. Giving your business card is customary, and try to have it printed in English and the local language as well. The main meal is at noon throughout Latin America. Most businesses close for two or three hours in the early afternoon because of this tradition, as well as the tropical heat. Proper business attire, even though it seems hot to you, is a jacket and tie for men, comparable attire for women.

When you arrive it is customary to give gifts. Appropriate business gifts are name brand items and perfume for women; for men name brand items or men's accessories; or for both, an item linked to the art or history of your own country or region. If your host invites you to a home for dinner, it is customary to bring flowers or good wine or liquor. Be prepared to follow with a toast of your own after your host says one first. Hospitality and generosity, by U.S. standards, is over the top. Figure 14.1 offers some general hints about business protocol in several countries in South America.

COUNTRY	GREETINGS	CONCEPT OF TIME	GIFT GIVING	CONVER- SATION
Argentina	Handshake and a nod	Appointments are required	Avoid personal items such as a tie	Discuss sports; avoid politics & religion
Bolivia	Handshakes are common	Visitors are expected to be punctual	Lunch or dinner in a restaurant is common	Attempts to speak Spanish are welcome; avoid politics and religion
Brazil	REMEMBER PORTUGUESE IS THE LAN-GUAGE	Customary to arrive ten min-utes late	If entertained in the home, send flowers and a note of thanks the next day to the hostess	Discuss chil-dren; avoid eth-nic jokes, poli-tics, Argentina, and religion
Chile	When first intro-duced, a hand-shake and a kiss on the right cheek is cus-tomary	Appointments are necessary; meetings start and end ON TIME	When visiting a home, flowers for the hostess are appropriate	Avoid politics and religion
Columbia	Handshakes are customary; use tile and last name. Discussion over coffee precedes any business transaction	Being on time is relatively important in the large cities.	Send fruit, flow-ers, or choco-lates before vis-iting a home	Discuss sports, art, the beauti-ful countryside; avoid politics
Ecuador	Give yourself time to get used to the alti-tude; hand-shake is cus-tomary	Stores close during the two-hour siesta	Lavish thanks are expected after being entertained in the home	Avoid politics

COUNTRY	GREETINGS	CONCEPT OF TIME	GIFT GIVING	CONVER-SATION
Paraguay	People stand very close when talking; wait until invited to use first names	Visitors are expected to be on time	Ask permission to enter when visiting a home	Discuss family, sports, current events, the weather; avoid politics
Peru	Handshakes on meeting and leaving	Only a bullfight requires punctuality	Flowers are always appropriate	Avoid local politics
Uruguay	A handshake is customary; first names used only by close friends	Meetings are formal, but start a few minutes late	Flowers or chocolates sent to the hostess if invited to a home	Sports are safe; avoid politics
Venezuela	A hug or handshake is common	Because they are busy, be prepared to be punctual and to get to the point directly	Usually only close friends get invited to their homes	

FIGURE 14.1 Common Business Protocols and Customs in Selected Latin American Countries

Source: Axtell, 1993

CHAPTER 15
NAFTA

Although the nations of North America have a distant history of conflict associated with the birth of nations, the United States, Mexico, and Canada have enjoyed friendly relations and non-hostile borders for most of this century. The United States and Canada have historically maintained close economic ties by engaging in trade across the border and investing in each other's economy. They signed the United States - Canada Free Trade Agreement in 1989.

Mexico followed economic policies under socialistic and nationalistic governments that was focused inward. Only in 1990 did Mexico consider a free trade agreement with the U.S. The change was part of Mexico's response to its debt crisis in 1982. U.S. investment in Mexico was stimulated by its Maquiladoras Program which granted tariff exemptions on imported equipment and material. The U.S. Harmonized Tariff Schedule allowed for U.S. made parts to reenter duty free after they had been processed into new products. Such exemptions provided incentive for the two nations to integrate economic activities and to increase co-production.

On December 17, 1992, the leaders of the United States, Canada, and Mexico signed the North American Free Trade Agreement commonly know as NAFTA. It set into motion the first trade pact between industrialized countries and a developing country. It also laid the structure for the largest free trade zone in the world with 360 million people and over $6 trillion in annual gross national product. Implementation of the agreement, expected gradually over a decade, began January 1, 1994.

NAFTA AND RELATED ORGANIZATIONS

NAFTA was a controversial treaty, so the three countries adopted related accords to address concerns about environment, labor, and border area development. A series of organizations — secretariats, commissions, working groups — are being created to implement the agreements.

NAFTA; agreements among the U.S., Canada, and Mexico; and agreements between only Mexico and the U.S. called for the creation of a number of organizations:

- Free Trade Commission supervises and implements NAFTA; it rotates meeting places in the three countries
- Commission for Environmental Cooperation, located in Montreal, is the result of the North American Agreement on Environmental Cooperation
- Commission for Labor Cooperation, located in Dallas, is the result of the North American Agreement on Labor Cooperation
- Border Environment Cooperation Commission, located in Ciudad Juarez, Mexico, is the result of a bilateral agreement between the United States and Mexico
- North American Development Bank, located in San Antonio, is the result of a bilateral agreement between the United States and Mexico. Plans also call for a Community Adjustment and Investment Program with offices in Los Angeles.

The three partners generally share equally the responsibility for establishing and managing the NAFTA commissions, boards, committees, and working groups. Leadership, location, staffing, and budgets are also shared equally. Since setting up these supporting bodies had no specific schedule, it has taken longer than expected. The trade ministers of the three countries agreed in principle to form the NAFTA Coordinating Secretariat to be located in Mexico City.

CANADA AS A NATION

Working in Canada, like working in the United Kingdom, can

pose a major obstacle for U.S. citizens. Common language, holidays, customs, cultures, beliefs can often lead you to the utterly false sense that you are among countrymen. YOU ARE NOT. YOU ARE THE FOREIGNER THERE. Like any other nation, Canadians are proud of their people, history, customs, and national symbols.

MEXICO'S CLIMATE FOR BUSINESS

Like the United States and Canada, Mexico is a nation that was born of European colonialism. Hernan Cortes defeated the Aztecs in 1521, and Mexico remained under Spanish rule until 1821 when independence was established. A rebellion in Texas in 1836, and a war with the United States forced Mexico to concede Texas, Arizona, New Mexico, and California to the U.S. by 1848. Modern Mexico is finally established in 1917 after a series of interventions by Spain, Great Britain, France, the U.S. and internal conflict.

Mexican history and culture shapes its complex population. Neither Spanish or Indian, Mexicans are a combination of ancient and contemporary, of traditional and stylish; a clear contrast to their northern neighbors. They are driven by spirit and emotion rather than consumed with efficiency, organization, and punctuality like their neighbors.

A Mexican's sense of time is not driven by the urgency to accomplish things before it is too late. They see birth and death not as a beginning and end, but they interpret them as part of the continuity of a living past. On their Day of the Dead, the day after Halloween, Mexicans go to the graves of their ancestors with flowers, food, and drink to celebrate the continuity of their lives.

In this context of time, planning is almost unnatural. Being on time may be considered rude, and ignoring an appointment is no occasion for offense. Absenteeism on Mondays is so common it is called "San Lunes." A failure or accident is often met with the response "ni modo," or no way it could be avoided.

Business in Mexico is still male dominated, and women there must deal with the strong sense of machismo. Status and appearances are important in Mexico, so titles and names are more formal. It is common to use a title that reflects education or company rank, as well as the traditional terms of eminence "Don" and "Dona." First names are used only if there is an invitation to do so. Business dress is conservative and formal in the cities.

A normal business day is from 9 a.m. to 6 p.m. with a two or three hour siesta in the afternoon. Most foreign companies are based in Mexico City. Initial contact must follow a formal protocol, a letter and a follow-up phone call. Often the Mexican is slow to respond, but be persistent. The first appointment is to establish a personal relationship, and is critical for any future relationship with the company. After making an appointment, arrive on time for the first meeting, however expect to wait 30 minutes or more for the meeting to take place.

Some current management practices such as team-centered work groups and quality circles need to be translated into the Mexican culture. Mexicans value loyalty to family and friends, shunning individual reward. Amend such "quality" program practices to meet the needs of the Mexicans.

After establishing a relationship, if you are honored with an invitation to a Mexican home, plan to arrive at least 20 minutes late, and bring a gift for the host or hostess. Also plan on a late dinner since the evening meal is routinely served at 9 or 10 p.m.

CHAPTER 16
The Pacific Rim

In 1997 Hong Kong returns to Chinese rule after almost a century of British influence. Think of that city as a locus for Eastern and Western managerial values. The diverse character of the city is a combination of the forces of national culture and business environment. Both have an impact on how an individual thinks and behaves. In understanding and working with people of the Pacific rim, keep in mind that the influence of national culture tends to perpetuate differences, while the global environment of standard business practices tends to emphasize similarities. (See Figure 10.2 above) In the Pacific, the national, ethnic, racial, cultural, social, and economic diversity of the nations, countries, and states offers challenges and opportunities for foreigners working there. Two cultures and nations predominate: Japan and China. Since Marco Polo made his overland journey there centuries ago, East-West relations have presented a continued challenge and opportunity, and each has influenced the other.

Hong Kong and Singapore are examples of what happens when East meets West. A sort of hybrid emerges, blending some values and beliefs that are the result of business practice, as well as some that are the result of national culture. Hong Kong, because it developed under British capitalism, has a business environment similar to the U.S. It also retains the strong cultural influences of Confucian tradition and Chinese customs. Figure 16.1 shows the dynamic pull of forces among Eastern and Western business cultures.

In order to understand the behavior of people in organizations, some of the socially desirable characteristics measured in each should shed some light. Some common measures in the West are Machiavellianism, dogmatism, locus of control, and tolerance of ambiguity. In the East Confucian dynamism, there is human-heartedness, integration, and moral discipline.

- High Machiavellianism scores on this measure of behavior indicate a preference to use power to meet a desired goal
- High Dogmatism scores indicate a more rigid personality
- High Locus of control scores here indicate a feeling of being controlled externally
- High Tolerance of ambiguity scores show a desire for more certainty
- Confucian dynamism, a philosophical value system based on the teachings of Confucius
- Human-heartedness is a concept similar to Hofstede's "masculinity," (see Chapter 2 and Figure 2.2)
- Integration is a concept similar to Hofstede's "power

CHINA ◄────► HONG KONG ◄──► UNITED STATES

| Collective Eastern Culture with roots in:
-Socialism
-Communism
-Confucianism
Technologies under-developed
No internal structure for world commerce | Recently emerged capitalistic state
British colonial influence on:
-Economy
-Education
-Legal system
98% Cantonese Speaking & follow traditional Cultural Patterns
Well developed financial system essential in world commerce
Link between China and the West
Base for overseas companies with ventures in China
Constantly in touch with both East and West | Western Culture Capitalist business environment
English and European influence on:
-Legal system
-Religion
-Political system
-Education
-Economy
-Social system
-Cultural values
Highly developed technologies and sciences |

FIGURE 16.1 Hong Kong Offers an Example of the Dynamic Forces at Work Between East and West

distance," (see Chapter 5) that is the relative closeness or distance between managers and subordinates
- Moral discipline is similar to Hofstede's "individualism," (see Chapter 2 and Figure 2.3) and has these five items: moderation, keeping oneself disinterested and pure, having few desires are positive, adaptability and prudence are negative.

CHARACTERISTIC	COUNTRY	low	med	high
WESTERN Machiavellianism	United States Hong Kong China	 x	 x 	x
dogmatism	United States Hong Kong China	x 	 x 	 x
locus of control	United States Hong Kong China	x 	 	 x x
tolerance of ambiguity	United States Hong Kong China	 x x		x
EASTERN Confucian dynamism	United States Hong Kong China	x 		 x x
human-heartedness	United States Hong Kong China	x 		 x x
integration	United States Hong Kong China	x 		 x x
moral discipline	United States Hong Kong China	x 		 x x

FIGURE 16.2 Business Values Among Eastern and Western Countries Differ Widely

Source: Journal of International Business Studies, 1993

Figure 16.2 compares responses among managers from the U.S., Hong Kong, and China to these values.

China, however, is awakened to the value of private industry. Unlike the Russian's explosive leap from communism to a market economy, China has slowly relaxed national control from many parts of its economy. "There are 25 million "joint ventures," "collectives," "shareholding companies," "township and village enterprises" and civilian enterprises" - euphemisms for firms that aren't socialist." (Kahn, Joseph. "Spreading Capitalism, New Entrepreneurs Are Remaking China," *The New York Times* July 20, 1995:A1.)

JAPANESE MANAGEMENT CULTURE

The Japanese management and business culture has had an enormous impact on the world economy, high technology manufacturing practices, and general concepts of quality business processes. In the United States and most of the European Union, capitalism rests on free markets, private ownership, investment, and competition. In the United States in particular, collective solutions to social problems are not usually favored. Government, unions, professional associations, and employer associations are not really expected to play a role in the running of a private enterprise. Increasing shareholder value is the common measure of performance. As noted in our discussion of the European Union, other groups in capitalistic countries often have a great deal of influence. This is particularly true in Japan.

> Capitalism in Japan is far less individualistic than its U.S. counterpart. But it is also focused largely on private gain, although it is the company or *keiretsu* that benefits from current earnings rather than dividend-receiving individual investors. Japanese managers are expected to use profits to fund high and sustained levels of corporate investment, not to distribute them to shareholders (who have traditionally made money on share-price appreciation) or to company employees. (*Harvard Business Review* July-August, 1992:59)

Working with Japanese businesses it is important to remember that Japan is a CIVILIZATION, NOT A MARKET. With this in mind consider that certain themes in Japanese history have shaped its business practices. The Japanese have a long history of assimilation and transformation of foreign ideas and techniques. Even though they emulate the Western world, they have a powerful sense of their cultural uniqueness. Their managers have an outstanding talent for organization, and they can plan and implement large-scale projects. They also have historically sought world-wide admiration.

The Japanese company should be considered a social entity, seeking economic goals. In that context, companies create a distinctive company spirit. Learning and self-improvement are driving forces in these organizations. The company becomes the embodiment and the transmitter of Japanese values.

The style of Japanese managers is one of mentorship. They are guides, philosophers, and friends to younger employees. They indirectly influence through intense informal pressure; some might call it coercion. Their power builds through the creation of networks, since the society is based on an elaborate system of exchange of favors and obligations. "Sincerity" is a highly valued trait. This is the capacity to behave in such a way so that you or your company won't be in trouble.

Business relations between the United States and Japan have always been highly competitive. Trade negotiations are always hard fought. Disputes over access to markets in construction, cellular phones, aviation routes, photographic film, prescription drugs, electronics, luxury cars, and software dispel the myth that "the Japanese are so adverse to confrontation, or so dependent on U.S. military protection that they would never challenge Washington." (Davis, Bob and Michael Williams. "U.S., Japan Usher in a Nastier New Era," *The Wall Street Journal* June 29,1995:A12) Since the end of the Cold War, trade confrontations between allies have become more common.

However, individual dealings must adhere to the extraordinary sense of politeness. The concept of saving face, as we discussed in Chapter 3, is extremely strong. Japanese go to extraordinary lengths to save face, so you should too. Business protocol is formal, and the Japanese are punctual in both business and social settings. They use their first name only for family and very close friends. Handshakes common in Hong Kong and China are not common in Japan. The usual greeting is a long and low bow. A business card is not given lightly or without a bit of ceremony. Always hand it to your counterpart with both hands.

Gift giving in Japan is an institution, art, and a revered custom. Their centuries-old customs spell out the type of gift to give, how it should be wrapped, and how it should be presented. The Japanese do not expect you to know all the customs. According to The Parker Pen Company, here are a few elements of the gift-giving style and form:

- Never surprise the receiver.
- Wrap the gift.
- Allow the receiver to open the gift later.
- Give and receive gifts with both hands.
- Comment on the modesty and insignificance of your gift.
- Never give four of anything (Shi for four is associated with death).
- The value of the gift befits the status of the recipient.
- Gifts are usually exchanged at the end of the visit.
- Try not to get caught empty-handed for an unexpected gift.
- Expect to receive a gift in return.
- In Japan the gift is an expression of the relationship.
- Style is more important than substance in Japan.

Since Asia and the Pacific Rim are so different from the West, welcome the challenge to learn and experience something new and exciting every day. As a foreigner you will always be an outsider to the culture you can never really comprehend. Your business dealings anywhere in the world with different cultures gives you insight into your own.

APPENDICIES

Further Reading

FURTHER READING

Adler, Gordon. "The Case of the Floundering Expatriate," *Harvard Business Review* July/August 1995:24-40.

Axtell, Roger. *Dos and Taboos Around the World* 3rd Ed. Compiled by The Parker Pen Company. NY: Wiley, 1993.

Background Notes. Washington, D.C.: Department of State. Annual.

Barbee, George and Mark Lutchen."Local Face, Global Body," *PW Review* Spring 1995: 18-31.

Barnlund, Dean C. *Communicative Styles of Japan and America: Images and Realities.* Belmont, CA: Wadsworth, 1989.

Business America: The Magazine of International Trade. Washington, D.C.: U.S. Department of Commerce. (Published biweekly.)

Copeland, Lennie & Lewis Griggs. *Going International: How to Make Friends and Deal Effectively in the Global Marketplace.* New York: Random House, 1985.

Culture Grams. Provo, UT: Center for International Studies, Brigham Young University. Set of 96 cultures.

Department of State Publication 7877. Washington, DC: 1993.

Destination Japan: A Business Guide for the 90's. Washington, DC: GPO, 1991.

Doyle, Edward. *How the United States Can Compete in the World Marketplace.* NY: IEEE, 1991.

Directory of American Firms Operating in Foreign Countries. NY: World Trade Academy Press. Annual.

Directory of Foreign Firms Operating in the United States. NY: World Trade Academy Press. Annual

Edwards, Mike. "A Broken Empire: After the Soviet Union's Collapse," *National Geographic* 183:3 (March 1993):2-53.

"Enterprise Funds," GAO/NSIAD-94-77. Washington, DC: 1994.

Europe: The Magazine of the European Community. Washington, DC.: EC Delegation to the United States. (10 times per year)

Europe: World Partner—The External Relations of the European Community. Luxembourg: Office for Official Publications of the European Communities, 1991.

The European Community 1992 and Beyond. Luxembourg: Office for Official Publications of the European Communities, 1991.

The European Community in the Nineties. Washington, DC.: EC Delegation to the United States, 1992.

European Union. Luxembourg: Office of Official Publications of the European Communities, 1994

Export Programs: A Business Directory of U.S. Government Services. Washington, D.C.: U.S. Department of Commerce, 1994.

Former Soviet Union. GAO/GGD-95-60. Washington D.C.:1995.

Ferraro, Gary. *The Cultural Dimension of International Business,* 2nd Ed. Englewood Cliffs, NJ: 1994.

Frederick, Howard. *Global Communication and International Relations.* Belmont, CA: Wadsworth Publishing Company, 1993.

Gannon, Martin. *Understanding Global Cultures.* Thousand Oaks, CA: Sage, 1994.

Goodman, Michael B. "The Special Section on Professional Communication in Russia: An American Perspective," *IEEE Transactions on Professional Communication* 37:2 June 1994:90-91.

Hall, Edward T. *Beyond Culture.* NY: Doubleday, 1976.

—————. *The Dance of Life.* Garden City, NY: Anchor/ Doubleday, 1987.

—————. *Hidden Differences: Doing Business with Japan.* Garden City, NY: Anchor/ Doubleday, 1987.

—————. *The Hidden Dimension.* NY: Doubleday, 1966.

—————. *The Silent Language.* NY: Doubleday, 1959.

Hall, Lynne. *Latecomer's Guide to the New Europe: Doing Business in Central Europe.* NY: American Management Association, 1992.

Haglund, E. "Japan: Cultural Considerations," *International Journal of Intercultural Relations* 8 (1984): 61-76.

Henzler, Herbert. "The New Era of Eurocapitalism," *Harvard Business Review* July-August 1992: 57-68.

Hodgson, Kent. "Adapting Ethical Decisions to a Global Marketplace," *Management Review* (May 1992): 53-57.

Hofstede, Geert. *Cultures and Organizations.* London: HarperCollins, 1991.

International Business Practices. Washington, DC: Commerce Department #003-009-00622-8)

Kohls, Robert. *Survival Kit for Overseas Living,* 2nd Ed. Yarmouth, Maine: Intercultural Press, 1984

Lawrence, Paul and Charalambos Vlachoutsicos. "Joint Ventures in Russia: Put the Locals in Charge," *Harvard Business Review* January-February 1993:44-54.

Moore, Geoffrey. *Crossing the Chasm: Marketing and Selling Technology Products to Mainstream Customers.* NY: Harper, 1991.

Noâl, Emile. *Working Together — The Institutions of the European Community.* Luxembourg: Office for Official Publications of the European Communities, 1994.

"North American Free Trade Agreement: Assessment of Major Issues." GAO/GGD-93-137 A & B. Washington, D.C.: 1993.

"North American Free Trade Agreement: Structure and Status of Implementing Organizations." GAO/GGD-95-10BR. Washington, D.C.: 1993.

Pagell, Ruth and Michael Halperin. *International Business Information: How to Find It, How to Use It.* Phoenix, AZ: ORYX, 1994.

Piet-Pelon, Nancy and Barbara Hornby. *Women's Guide to Overseas Living.* [PLACE: PUB, Date].

Rowland, D. *Japanese Business Etiquette: A Practical Guide to Success in the Global Market Place.* New York: Praeger, 1986.

Semler, Richardo. "Who Needs Bosses?" *Across the Board*
(February 1994):

Terpstra, V., and K. David. *The Cultural Environment of
International Business.* Cincinnati: South Western, 1985.

*Tomorrow's Company: The Role of Business in a Changing
World.* London: RSA (Royal Society for the encouragement of Arts, Manufactures & Commerce), 1994.

Trompenaars, Fons. *Riding the Waves of Culture:
Understanding Cultural Diversity in Business.* London:
The Economist Books, 1993.

Victor, David. *International Business Communication.*
New York: HarperCollins Publishers, 1992.

Weiss, Stephen. "Negotiating with 'Romans'," *Sloan
Management Review,* Winter 1994:51-61.

INFORMATION RESOURCES

The European Union: Washington, D.C. 1-(202)-862-9500;
New York City 1-(212)-371-3804.

U.S. Commerce Department, International Trade
Administration, Trade Information Center; 1-800-USA-
TRADE (1-800-872-8723)

U.S. Departments of State and Commerce, Country Desk
Officers; 1-(202)-482-3022; For a specific country desk
officer - 1-(202)-647-4000.

U.S. Department of State Coordinator for Business Affairs;
1-(202)-647-1942; FAX (202) 647-5713.

APPENDIX 1
International Business Ethics

Working internationally places you under both United States and foreign laws and regulations. Most corporations have an ethics code with a section on international business ethics. *The Westinghouse Code of Ethics & Conduct* (1994) offers a fine model:

> Employees conducting business internationally are required to comply with all applicable U.S. and foreign laws and regulations. Compliance with such laws, as well as company standards (including this Ethics Code), is required even if they seem inconsistent with local practice in foreign countries, or would place the company at a competitive disadvantage. The penalties for non-compliance can be severe, both for the company and for involved individual employees.

KEYS TO COMPLIANCE

Don't Make or Offer Unlawful Payments or Bribes	*The Foreign Corrupt Practices Act* bars the payment or offering of anything of value to officials or politicians of foreign governments, and others, to obtain or retain business. It also requires proper accounting for transactions. The company has a specific policy concerning retention of overseas sales agents.
Abide by Import/Export Controls	A number of U.S. government controls restrict, to varying degrees, the import and export of goods, services, and technical information to various countries, as well as the re-export of U.S. products from other countries. Foreign countries may have similar laws that apply to U.S. products. Employees must comply with these laws applied to their businesses and products, and specifically by obtaining the necessary general or validated import/export licenses.
Adhere To U.S. Economic Boycott Laws	U.S. laws restrict trading with certain foreign countries, and prohibit U.S. companies from complying with certain boycotts imposed by other countries. These laws cover U.S. persons and can also apply to Westinghouse subsidiaries located outside the U.S. Anti-boycott regulations also require notification to the U.S. government of any boycott request received from a foreign government or official. Boycott laws, including the countries affected, often change and must be closely monitored.
Refer International Trade Law Questions to the Law Department	The application of U.S. and foreign laws can be very complex. Sometimes, U.S. laws conflict with the laws of other countries. When such conflicts appear in the conduct of your business, contact the Law Department.

APPENDIX 2
Local Presence, Global Concerns

Here is an Action Plan for a local/global strategy from Price Waterhouse executives George Barbee and Mark Lutchen ("Local Face, Global Body," *PW Review* Spring 1995: 18-31.)

- Cultivate Key Relationships
- Observe Local Protocol
- Listen
- Partner with Stakeholder and Make Time to Build with Care
- Be Sensitive
- Communicate
- Organize Multidimensional Teams
- Assess Often
- Be Consistent
- Learn, Master, and Use Best Practices

APPENDIX 3
Cultural Metaphors

COUNTRY	CULTURAL METAPHOR	CHARACTERISTICS REVEALED BY THE METAPHOR
England	Traditional British House	Laying the Foundations - History, Politics, Economics Building the Brick House - Growing Up British Living in the Brick House - Being British
Italy	The Italian Opera	Pageantry and Spectacle Voice to Express Words in Music Exteriority, the Belief that emotions are So Powerful that an Individual Cannot Keep Them Within and Must Express Them to Others Chorus (unity of the culture) and Soloists (regional variations)
Germany	The Symphony	Orchestra, like Society, Made of Individuals Individuals Subordinate to the Greater Good, Music (society) Individuals defer to the wants of the Conductor and the Needs of the Symphony Everyone Cannot be a Soloist, or Wants to Be; Soloists Time is Brief Discipline of the Individuals Allows the Symphony to Flower and Flourish
France	French Wine	Purity Classification Composition Suitability The Maturation Process
Russia	The Ballet	Echelons of the Ballet Theatrics and Realism The Russian Soul - complexity of Russian culture, the grand expression of aristocracy and the gentle beauty of the countryside
Ireland	Conversation	Intersection of Gaelic and English Prayer as Conversation Free-Flowing Conversation: Irish Hospitality Places of Conversations: Irish Friends and Families Ending a Conversation
Turkey	The Coffeehouse	Islam and Secularity Recreation, Communication, and Community Integration A Male Domain A Modest Environment Life Outside the Coffeehouse

COUNTRY	CULTURAL METAPHOR	CHARACTERISTICS REVEALED BY THE METAPHOR
Israel	The Kibbutzim	Community Ownership of All Property Absolute Equality of Members Democratic Decision-making Value of Work as Both End and Means Communal Responsibility for Child Care Primacy of the Group Over Individuals
Japan	The Japanese Garden	Wa - harmony Shikata - the proper way of doing things with emphasis on the form and order of the process Seishin, or "Spirit" Training Combining Droplets or Energies Aesthetics

India	The Dance of Shiva	The Dance Symbolizes:
		  srishi - creation & development
		  sthiti - preservation & support
		  samhara - change & destruction
		  tirobhava - shrouding, symbolism, illusion, & giving rest
		  anugraha - release, salvation, & grace
		Cyclical Hindu Philosophy
		The Cycle of Life
		The Family Cycle
		The Cycle of Social Interaction
		The Work and Recreation (Rejuvenation) Cycle

United States	American Football	Individualism and Competitive Specialization Huddling Ceremonial Celebration of Perfection
China	The Chinese Family Alter	Confucianism and Taoism Roundness, symbol of family continuity and structural completeness Harmony within the family and the broader society Fluidity, capacity to change while maintaining traditions

FIGURE A3 Often an Event or a Practice in a Country or Society Offers a Metaphor for Its Culture, Revealing Some Fundamental Values and Beliefs

Source: Gannon, Martin. Understanding Global Cultures (1994)

APPENDIX 4
National Cultures and Corporate Culture Types

International Corporate Culture Types:
- The Family
- The Eiffel Tower
- The Guided Missile
- The Incubator

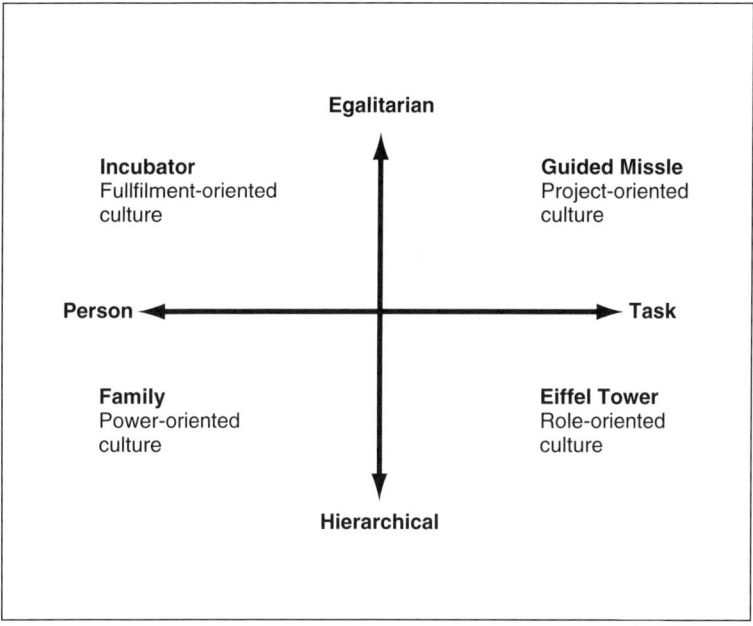

FIGURE A4 The Four Corporate Culture Types Show: 1) The Relationship of Employees to the Organization; 2) The System of Authority Defining Superiors and Subordinates; 3) The Employee's View of the Organization's Destiny, Purpose and Goals.

Source: Fons Trompenaars. Riding the Waves of Culture (1993): 138-141

APPENDIX 5
Gathering Facts for a County Analysis

COUNTRY OR REGION:

1 **National Language** Local Languages Dialects **1A National Symbols** Flag, Songs, Colors, Flowers, Holidays	1. 1A.
2. **National Foods**	2.
3. **Politics and Government** System of Government Political Parties Organization of Government Political Leaders Police System Military	3.

4. **Economics & Industry**
Major Industries
Imports/Exports
Foreign Investment
Industrial Development
Agriculture
Fishing
Markets
Urban & Rural Conditions
4A. Resources, Communication
    & Transportation
Human
Natural - Climate &
   Geography; Minerals
Agriculture, Forests, Water
Comm. Infrastructure
Media - Print & Broadcast
Computer Network
Trans. Infrastructure
	4.
	
	4A.

5. Arts and Culture Painting & Sculpture Music & Dance Literature, Drama, Poetry TV & Radio Movies & Cinema Architecture Folk Arts, Crafts	5.
6. Education Educational Philosophy School System Colleges & Universities Vocational Training	6.
7. Science Inventions & Achievements Technological Infrastructure Attitude toward Science Institutions & Laboratories Medicine & Hospitals	7.
8. Religion & Philosophy Modern Beliefs Folk Tales & Superstitions Sayings & Proverbs Myths & Legends	8.
9. Family & Social Structure Customs & Rituals - birth, death, marriage Social Welfare Systems	9.
10. Sports & Games Local Sports Children's Games Contemporary World Sports	10.

FIGURE A5 Checklist for Gathering Basic Factual Information About the Country or Region Where You Plan to Work

APPENDIX 6
What to Bring — Logistics

ITEM/AREA	ACTION
Documents	Apply for passport and visas Get International Driver's License
Medical	Have a medical exam before you leave, and any immuniza- tions necessary, including hepatitis Obtain copies of records, x-rays, prescriptions Know blood type; shots Obtain prescriptions for glasses Have a dental exam; get dental records See a vet for appropriate shots and certificates for your pet.
Legal	Update your will before you go Designate power of attorney to a responsible relative or friend
Financial	Buy local currency for countries you will be going through for transportation, tips, etc. Buy traveler's checks to cover travel expenses Arrange financial transfer and access with your bank - mailing statements, access to safe deposit boxes, power of attorney authorization and signature cards Lines of Credit and Credit Card Notification Arrange With An Accountant to Have Proper State and Federal Tax Forms Filed Meet with Your Insurance Agent to Discuss Medical, Life, Home, Auto, Fire, Accident Coverage at Home While You are Away; Discuss Coverage for the Country You Will Be working In Keep Receipts of All Expenses Related to Your Move
Education	Meet with Teachers and Administrators to Discuss Tests, evalu- ations, Transfer of Records, Placement in Schools in the Host Country Contact Schools in the Host City Well in Advance; Select the School and Arrange for Spaces for Your Children
Communication	Get a Change of Address Kit From the Post Office Notify Family Members of Your New Address and the Customs Requirements Stop Deliveries of Newspapers, Magazines, and Other Services Arrange for Mail and Telephone Service Where You Plan to Be
Transportation	Visit AAA for an International Driver's License
Household	Notify Your Utility Services of Your Plans, and Arrange to Discontinue Service - Telephone, Gas, Oil, Water, Electricity

FIGURE A6. Checklist to Start Planning for Your Overseas
Assignment

Source: Robert Kohls, Survival Kit for Overseas Living, 2nd Ed

APPENDIX 7
Doing Business Overseas —
Whom to Contact

Before you go, become familiar with the official U.S. presence in the host country, and the structure of a typical U.S. Diplomatic Mission. Also, find out specific contacts from the Departments of State and Commerce:

- U.S. Commerce Department, International Trade Administration, Trade Information Center; 1-800-USA-TRADE (1-800-872-8723)
- U.S. Departments of State and Commerce, Country Desk Officers; 1-(202)-482-3022; For a specific country desk officer - 1-(202)-647-4000.
- U.S. Department of State Coordinator for Business Affairs; 1-(202)-647-1942; FAX (202) 647-5713.

KEY OFFICERS	RESPONSIBILITIES
Chief of Mission - Ambassador, Minister, Charge d'Affaires	All components of the U.S. Mission within a country, including consular posts
Commercial Officer	Assists U.S. business through: arranging appointments with local business and government officials; counseling on local trade regulations, laws, and customs; identifying importers, buyers, agents, distributors, and joint venture partners for U.S. firms; and other business assistance
Commercial Officers for Tourism	Implement marketing programs to expand inbound tourism; increase the export competitiveness of U.S. travel companies, and strengthen the international trade position of the U.S.
Economic Officers	Analyze and report on macroeconomic trends and trade policies and their implications for U.S. policies and programs
Political Officers	Analyze and report on major financial developments
Financial Attaches	Analyze and report on political developments and their potential impact on U.S. interests
Labor Officers	Follow the activities of labor organizations to supply such information as wages, non-wage costs, social security regulations, labor attitudes toward American investments, etc.

KEY OFFICERS	RESPONSIBILITIES
Consular Officers	Extend to U.S. citizens and their property abroad the protection of the U.S. Government. Maintains lists of attorneys; acts as liaison with police and other officials; has authority to notarize documents. The State Department recommends that business representatives residing overseas register with the consular officer; in troubled areas, even travelers are advised to register
Administrative Officers	Responsible for normal business operations of the post, including purchasing for the post and its commissary
Regional Security Officers	Responsible for providing physical, procedural, and personnel security services to U.S. diplomatic facilities and personnel; responsibilities extend to providing in-country security briefings and threat assessments to business executives
Security Assistance Officers	Responsible for Defense Cooperation in Armaments and foreign military sales to include functioning as primary in-country point of contact for U.S. Defense Industry
Scientific Attaches	Follow scientific and technological developments in the country
Agricultural Officers	Promote the export of U.S. agricultural products and report on agricultural production and market developments in the country
AID Mission Officers	Responsible for AID Programs, including dollar and local currency loans, grants, and technical assistance
Public Affairs Officer	Press and cultural affairs specialists; maintain close contact with local press
Legal Attaches	Serve as representatives to the U.S. Department of Justice on criminal matters
Communications Programs Officers	Responsible for the telecommunications, telephone, radio, and diplomatic pouches, and records management programs within the diplomatic mission. Maintain close contact with the host government's information/communications authorities on operational matters
Information Systems Managers	Responsible for the post's unclassified information systems, database management, programming, and operational needs. Provide liaison with appropriate commercial contacts in the information field to enhance the post's systems integrity
Animal and Plant Health Inspection Service Officers	Responsible for animal and plant health issues as they impact U.S. trade and in protecting U.S. agriculture from foreign pests and diseases. Expedite U.S. exports in the area of technical sanitary and phytosanitary (S&P) regulations

FIGURE A7. U.S. Mission Key Officers and Their
Responsibilities: A Wide Range of Help and
Information for U.S. Citizens In-country

Source: Department of State Publication 7877 (1993)